DATE DUE

NOV 0 9 1993			
DEC 2 2 2002			

D1088328

SCOPE OF
EXPERIMENTAL
ANALYSIS

SCOPE OF EXPERIMENTAL ANALYSIS

HUDY C. HEWITT, JR.

Department of Mechanical Engineering
Tennessee Technological University

PRENTICE-HALL, INC.

Englewood Cliffs, New Jersey

Library of Congress Cataloging in Publication Data

Hewitt, Hudy C
    Scope of experimental analysis.

    Includes bibliographical references.
    1. Engineering—Statistical methods. 2. Experimental design. I. Title. II. Title: Experimental analysis.
TA340. H45   620′.0042   72–8869
ISBN 0–13–796649–0

10 9 8 7 6 5 4 3 2 1

Printed in the United States of America

Prentice-Hall International, Inc., *London*
Prentice-Hall of Australia, Pty. Ltd., *Sydney*
Prentice-Hall of Canada, Ltd., *Toronto*
Prentice-Hall of India Private Limited, *New Delhi*
Prentice-Hall of Japan, Inc., *Tokyo*

CONTENTS

PREFACE

The book has been written for the undergraduate engineering student in an effort to present a balanced approach to the solution of engineering-type problems. It is particularly important that the mathematical and analytical tools that have been developed in the student be put to optimal use. Chapters one, two, three, and four introduce the fundamentals of the experimental methods of problem solving. The interface between analytical knowledge, experimental program design, and the measurement system are stressed.

Most of the material in these first four chapters is presented to provide a foundation for the general problem solution. Most of the details involved in designing an experiment must be developed by either reading references or practical laboratory experience. The role of the measurement system in this process is presented and the basic fundamentals involved in the measurement process are discussed.

Chapter five presents some of the primary techniques commonly used in the analysis of experimental data. The interrelationships that exist between the problem to be solved, the experimental program, and the measurement system are partially controlled by the final data analysis required. This chapter is not essential to the understanding of the rest of the book, but it is important to an understanding of what is required of a measurement system, of the control parameters, and of the types of solutions that are available.

Chapter six presents the foundations of measurement and the details of the measurement system. These details are continued throughout the remaining chapters. It is hoped that a logical approach to the measurement problem can be developed and that the basic concepts of measurement can be applied to any type of measurement problem.

The author gratefully acknowledges the influence and assistance of those who made helpful suggestions in the completion of this book. Professor Warren M. Rohsenow took the time to read the first draft and offer encouragement in this effort. Professor E. M. Sparrow read the second draft and made many very helpful comments on both organization and content. Professor John P. Wallace (my late colleague at Tennessee Technological University) taught a course in measurements using these notes and made some valuable comments. The author is also grateful to his wife, Mary Alice, for proof-reading the final draft and to the many secretaries on the Mechanical Engineering staff who contributed time and effort in typing the original and two later drafts.

HUDY CREEL HEWITT, JR.

1
THE SCOPE OF
EXPERIMENTAL ANALYSIS

1.1
Introduction

Experimental studies have provided the momentum that has resulted in most of the major scientific developments throughout history. The "Dark Ages" were, in part, a result of the philosophy propagated by the great thinkers of the "Greek Era," who based the fundamentals of science on abstract thinking. For them, it was unnecessary to prove a physical law by experimentation, since a proof based on logical deductions was the supreme test. The scientific laws formulated in this era of abstract thought became so firmly accepted that anyone able to show an experimental violation of these principles was thought to be a demon or witch. This blind faith in logic, with no provision for experimental proof, could not prevail in a truly scientific society.

Although there must have been many scientists who silently disagreed with a totally abstract approach to science, Galileo was one of the first to demonstrate its fallacies publicly. His experiment, proving that two spheres of different weight fall with the same velocity, was rejected; he was sentenced to die for his efforts. Only by renouncing the experiment as fraudulent was his life spared. Newton's (perhaps hypothetical) observation of an apple falling from a tree was the seed that grew into the statement of gravitational attraction. Kepler's laws of planetary motion were based on Brahe's experimental data of planetary motion. In general, the development of experimental techniques led to the acceptance of the present sceintific method of problems solving.

Many of the most honored scientists of modern times made their greatest contributions based on experimental results. Madame Curie discovered

1

radiation experimentally. Niels Bohr's model of the atom resulted from observed scattering of particles. Einstein's theory of relativity resulted from the Michelson–Morley experiment. Alexander Graham Bell, Louis Pasteur, the Wright Brothers, and Thomas A. Edison were all outstanding experimentalists. These were the pioneers of the experimental era, in which technical knowledge was secondary to performance.

The list of contributors to the development of experimental analysis could not be contained in one book. The efforts of these contributors have resulted in the acceptance of experimental verification as an integral part of the scientific method. An abstract hypothesis is no longer accepted as a scientific law until the hypothesis has been validated by experimental proof. According to Einstein, no amount of experimental verification can prove a law; but a single violation shown experimentally can disprove a law.

In this age of multimillion-dollar space probes, the value of rigorous experimental development has been adequately demonstrated. Most engineering graduates become involved with experimental programs in their initial positions with industry. Therefore, the importance of a complete understanding of experimental analysis cannot be overemphasized. All of the technical knowledge gained in school or by experience can be profitable only if a workable product results. This product is a balanced problem-solving capability.

1.2 What is Experimental Analysis?

Experimental analysis is a program in which physical phenomena are evaluated. The program may be as simple as measuring the dimensions a refrigerator with a ruler to see if it will fit in a given room space. The program could also be the development of a spaceship to land on Mars. In an attempt to encompass all programs into the framework of one subject, it is advantageous to consider experimental analysis in three parts:

1. Design and execution of the experiment
2. Design of the measurement system
3. Analysis of the program

The rest of this text is devoted to these three topics. Therefore, only a very brief description will be presented here.

The design of the experiment includes defining the problem in detail, collecting all available technical knowledge on the problem, selecting an experimental program, determining the parameters to be controlled, and executing the experimental program. All five steps are inherently interrelated and should be followed to develop the technique of problem solving.

Defining the problem is important so that an analytical and experimental program can be complete in its formulation. This also allows the literature review to span the subject matter that is necessary for the problem solution. Selecting the experimental program may involve scale modeling, analog modeling, or prototype testing. Control parameters are determined from the physical laws governing the phenomena being investigated. Actual performance of the experiment will introduce feedback into the other steps. Once the execution of an experiment has started, the information gained can be used to augment the original design of the experiment.

If the physical phenomenon under consideration can be represented by a mathematical model, the basic experiment may involve the use of either a digital or an analog computer to design a system. The computer allows for consideration of a range of control variables with relative ease. It is possible to predict the effect of many combinations of input variables in a relatively short time. For many problems computer design will be much less expensive and will provide much more information than could be obtained by building an experimental model.

The design of the measurement system involves determining the variables to be measured, evaluating the capabilities of various measurement systems, and selecting the proper system. Much of this text was prepared to introduce the principles of measurement in such a way that the readers will be qualified to design the desired measurement system. Here the emphasis will be placed on engineering-type measurement systems. The theory of measurements must be developed before engineers can be expected to design and use a measurement system accurately.

Finally, an analysis of the program provides a way to evaluate the credibility of the basic design of the experiment, consider error analysis of measurement data, and determine the success or failure of the program. Basic assumptions made in the original experiment design must receive critical review. The reliability of the measurement data must be determined. After the final analysis, the findings of the experimental program will usually be reported in the form of a written report. The scientific format for these reports is presented in Appendix I. This reference or a similar acceptable one should always be used in reporting findings.

1.3
Relative Importance of Experimental Analysis

The purpose of an engineering education is to prepare the individual to solve new engineering problems. To accomplish this the individual must know the fundamental physical laws and be able to apply them to new situations. A very broad background in physical laws is available in the many physics and engineering courses taken by the student. Mathematics courses equip the student with the methods for solving the mathematical models of the physical laws. However, all of the physical laws are based on experimental verification, and the purpose of this text is to present the experimental methods available for evaluating a hypothesis or design. In each subject the learner is provided with the fundamentals along with a step-by-step problem-solving technique. Often these techniques can be directly applied to different subjects, provided the user is qualified to think clearly in the different subject.

Of all the courses taken by students, only the laboratory and measurement courses are normally directed toward the development of knowledge in the area of experimental analysis. With the ever-increasing use of experimental design in problem solving, the importance of understanding the fundamentals of experimental analysis is obvious. It would be a misrepresentation to state that the day of the abstract thinker is gone, but his work must be accompanied by experimental information for his contributions to be completely realized. Even when the scientist does not actually perform the experimental program, he should understand the fundamentals of experimentation if he is to provide direction and leadership in the development of his ideas. Therefore, analytical and experimental methods are equally important, and the ideal balance of knowledge will include a fundamental understanding of both. For this reason it is extremely important that experimental analysis be presented in basic laws if parallels to analytical knowledge are to be possible.

The outstanding engineer must possess both technical knowledge of the physical laws and experimental expertise. A good understanding of one develops capabilities in the other. It is not possible to separate technical and experimental knowledge, since they do complement each other. Two purposes of this book are to present a logical approach to experimental analysis and to develop the basic concepts of measurement. This will allow the reader to glimpse the overall picture and to understand the role played by each part of his education in the total field of problem solving.

2
DESIGN OF
THE EXPERIMENT

The first step in experimental analysis is the design of the experiment. The success or failure of any program can often be determined by the foresight used in establishing the initial formulation of the problem being investigated. If you wish to leave Oklahoma City and drive by automobile to Los Angeles, there are many roads available. However, if you started driving toward New York, your chances of reaching Los Angeles are remote unless there is some change in direction. An analogous situation exists in the selection of the proper course of action to be pursued in the initial design of the experiment. The attainment of an experimental goal may be possible even if the initial approach is in error, but a well-thought-out initial approach results in a much more rapid solution of the problem.

The steps to be followed in solving the simplest problem should be the same as those followed for the most difficult ones. This kind of practice leads to a natural step-by-step problem-solving technique. Therefore, control-system problems, design of a product, operational procedures, and basic research can all be approached using a comfortable, effective technique. Four aspects of designing the experiment should become part of the problem-solving structure. These aspects will be discussed in the present chapter.

To be effective in any situation, the problem being considered must be clearly defined. Time spent in this effort is never wasted, because the uninformed scientist cannot make a contribution to the solution of a problem not understood. Once the problem is well defined, a complete search

5

of the technical literature should be made to develop a good technical background in the problem area. Sources for technical background material can be used to complete a literature review. Then, an experimental program can be selected that should be properly oriented. Finally, the parameters or variables, to be controlled, must be considered. Success in these four areas improves the chance of a successful program by providing a good initial course selection.

2.2
Problem Statement

An individual capable of recognizing problems and communicating these problems to people with the technical knowledge to solve them is indispensable. The mechanization of agricultural equipment proceeded rapidly because of adequate problem statements. An incomplete problem statement almost always results in an unsatisfactory solution.

The complete problem statement includes an overall statement of the problem to be solved, a statement of the environmental conditions, a statement of the developmental and product cost, a statement of the operating duration, and a statement of any pertinent facts associated with the problem. In terms of a systems approach to the problem, it is necessary to know the interactions of the system being considered with all outside systems. These statements must be made or understood before any attempt is made to solve the overall problem. Consider the problem statement: Build a rocket engine capable of developing a 20,000-lb thrust. Now suppose the engine was designed to be partially air cooled, and the engine was intended to power a space probe. The result would probably be an overheating failure.

The overall problem statement must be explicit. It should define the ultimate goal sought and the projected limitations acceptable. Understanding the overall problem is essential if maximum effort toward its solution is to be realized. Any work without an ultimate goal is usually wasted.

A statement of the environmental conditions may be a simple statement of where the problem is to be solved. Some examples of the variety of environmental conditions to be considered are:

1. The terrain of operation
2. The temperature variations
3. The wind conditions
4. The chemical environment
5. The humidity conditions
6. The pressure environment

7. Sunlight conditions
8. Vibrations and noise sources
9. The pollution problems

The list is by no means complete, but each of the above and other environmental conditions could change the method required to solve the overall problem.

A statement of the developmental and product-cost limitations has become a big factor in determining the feasibility of attempting to solve a particular problem. Careful evaluation of costs helps avoid termination of an investigation before completion. In industry, the balance of expected profits versus expended funds determines cost limitations. Often, basic research has no figure for expected profits and, therefore, cost support is based on a yearly budget. In any case the cost limitation should be known if an investigation is to proceed to a satisfactory conclusion. This information is usually not expected from the engineer, but he should seek an input from the proper source in order to contribute to the best solution of the problem.

A statement of the operating duration is the time limitations on either the development stage or the product stage. Time limitations on the development stage are for the purpose of planning the manpower to be expended. Time limitations on the product stage include estimated life of the product and duration of an operational period. It is good practice to forecast completion dates for various stages of the problem to avoid excessive time expenditures and poor time management. The development of a satsifactory product can be accomplished only if the desired operating conditions and operating life are known.

A statement of any pertinent facts associated with the problem is a method of obtaining any additional information that might influence the problem solution. In designing a product for the public, it might be important to know that the public does not like red air conditioners. This statement gives insight to specific problem conditions.

Looking only at the system to be analyzed, it is necessary to know all inputs to the system and all outputs expected of the system. All of this information should be contained in the definition of the problem. At this point it is possible to assess the general problem area and to dissect this problem more easily into its component parts. Then the system being considered can be divided into many systems for solution by small groups of scientists skilled in the particular area. Often the solution to one or more of the subsystems is not available in a usable form, and it is necessary to do a literature review of the work done by others in related areas. This leads to the second step in the design of the experiment.

2.3 Assimilation of Technical Knowledge

It is possible to solve a problem more easily and rapidly after the technical knowledge in the area of the problem has been developed to the extent that physical unkowns have been eliminated. Most research begins with a review of the literature associated with the problem being studied. The purpose of this review is to amass all of the technical knowledge presently available in a field and to point out areas where the physical governing laws are not well undertsood. If a problem has already been solved satisfactorily, it is usually advantageous to use this knowledge for a starting point. Since an engineer cannot have a broad enough background to work on any problem given him, he must learn how to develop his background in the areas related to the problem he is to solve.

Every successful engineer must be able to search through the literature and teach himself the fundamental relationships governing the problem with which he is working. Formal education in universities must, of necessity, be limited to the basic fundamentals of a relatively meager variety. New situations are often approached by the modification of existing technical knowledge. An assessment of the state of the art can be made after all previous work has been reviewed.

To develop the ability to research an unfamiliar problem area (or any other subject of interest), each reader should select a subject from Problem 2.1 at the end of this chapter and determine a list of publications for the technical knowledge on this subject. For example, if a problem is to determine the heat transfer away from a panel of electronic equipment in a spaceship, it would first be necessary to determine the mechanisms by which heat could be transferred. For space problems this might be only by radiation and conduction. Then, it would be necessary to determine the physical laws governing these two mechanisms of heat transfer: the Stefan–Boltzmann-type equation, and the Fourier-type equation. Finally, it would be important to search for any information concerning other work on a similar heat-transfer problem. This collection of information would be adequate for continuing the process of designing the experiment. However, for less familiar problem areas the process of reviewing the literature would be more extensive, since an understanding of the physics of the problem would have to be developed before any review could be meaningful.

A complete review of the literature can be accomplished in a very short time when a good library is available. A collection of articles published in the last year or so contains a bibliography of the work done previous to the paper being read. This very rapidly pyramids into a large collection of pertinent papers on the subject of interest. Therefore, finding one paper related to the subject leads to an entire list of papers related to this subject. Techniques for collecting a bibliography of papers concerning the problem being researched can be developed very rapidly by remembering that the most recent developments in a given area cannot be found in a textbook but must be found in the recent periodicals. Each reader should become aware of this by personally selecting a subject and looking at the publications.

Once all of the available technical knowledge has been collected, insight into the problem areas remaining and the difficulties experienced by other investigators allow for intelligent progress to a plausible problem approach. Even when the exact details of the problems to be solved do not fit any previous work, a background of technical knowledge in the problem area can be developed from a good literature review.

2.4
Selecting the
Experimental
Program

One of the most difficult decisions to make in the design of an experiment is that of selecting a model that adequately represents the problem being considered. Every mathematical equation representing a physical law is only a mathematical model for that law and is valid only as long as the physical law can be represented by that model. The equation

$$S = \epsilon E$$
$$(\text{stress}) = (\text{strain})(\text{modules of elasticity})$$
(2.1)

is valid for the stress–strain relationship for a material when the stress is below the elastic limit. However, this equation is not valid for any material having a parabolic or higher-order relationship between stress and strain. It is not valid when permanent set occurs or when creep is occurring. The equation cannot represent fatigue problems, and it is obviously not valid above the yield point of a material. Therefore, this mathematical model is valid only over a very restricted range of values of stress and strain. It should not be used outside this range, since it does not represent the physical phenomenon outside this range.

Almost every mathematical model of a physical phenomenon is valid over some restricted range and for some restricted boundary conditions. In selecting and using a model, all limitations must be involved in the final choice. This same principle is equally apparent in the selection of an experimental model to represent a given problem. A good understanding of the basic technical knowledge available on a given subject points to areas where additional knowledge must be obtained. To study these problem areas experimentally, either a form of experimental model or an analogous model must be constructed. If the mathematical equations governing a physical phenomenon are known, an experimental computer study can be used to evaluate the functional relationship between input and output parameters.

Two basic types of models are commonly used in engineering experimentation: scaled (or prototype) models, and analog-type models. The entire theory of modeling is based on the technique of constructing a system that responds to all inputs in the same way the system represented would respond. For example, the problem might be to determine the strength of a 6-in.-diameter steel shaft in tension. The tremendous loads required to accomplish this solution using the 6-in.-diameter shaft almost preclude testing the full-scale (or prototype) model. If the only input to be investigated is the axial loading, it might be possible to test a small-scale model of $\frac{1}{2}$-in.-diameter and predict the strength of the prototype based on this analysis. Since both models obey the relationship $S = \epsilon E$ for static loading, a strength test on the one-twelfth scale model could be used to predict the strength of the prototype. However, if additional inputs must be considered (i.e., end loading, transient environmental temperature, transient loading, vibration, and hysteresis), more elaborate testing of the scaled model must be considered.

It would be extremely pretentious to expect a simple spring–mass–damper to represent the motion of a limb of a tree in a windstorm. The spring–mass–damper system, $m\ddot{x} + c\dot{x} + kx = F$, might represent the tree limb if all coefficients for the tree were constant. However, the dampening, c, and the spring constant, k, for the limb would vary with the effective position of leaves and the liquid content of the limb. This type of modeling would require a much more detailed analysis of the tree and forcing function, wind. Therefore, an analogy between a model and the system it is to represent must include any and all variations of the true system from the model.

Actually, this fact makes the selection of the experiment much more interesting. It is not always possible to represent a real physical phenomenon completely with a simple model. A talent for experimentally modeling a system must be developed by continuing attempts to resolve problems. Confrontations where problems are not solved by simple analogies are the basic stimuli; they provide the impetus to develop the capabilities of the investigator.

One method of obtaining a model with the proper response to inputs is to ensure that the model has the identical governing equations as the prototype. This method also leads to analog-type modeling. A common example of this is the spring–mass–damper mechanical system and the inductance–capacitance–resistance electrical system. The equations governing these systems are

$$\text{mechanical:} \qquad M\ddot{x} + \bar{c}\dot{x} \times kx = F$$

$$\text{electrical:} \qquad L\ddot{q} + R\dot{q} + \frac{q}{c} = V \qquad (2.2)$$

Here the analogies are

$$\text{mass, } M \leftrightarrow \text{inductance, } L$$
$$\text{damping, } \bar{c} \leftrightarrow \text{resistance, } R$$
$$\text{spring constant, } K \leftrightarrow \frac{1}{\text{capacitance}} = \frac{1}{c}$$
$$\text{force, } F \leftrightarrow \text{voltage, } V$$
$$\text{displacement, } x \leftrightarrow \text{charge, } q$$

The influence of extraneous inputs, such as temperature and heat transfer, on these systems is often difficult to resolve. Determining the effects of various inputs is a major problem in any type of modeling.

When the problem being investigated can apparently be represented by some mathematical equation, the design of the experiment can be greatly enhanced by direct experimental verification using either a scaled or prototype model This proof would essentially provide any information required in the particular problem area. Once the differential equation that governs a physical phenomenon is known, many forms of modeling become available, including all types of similarity modeling, digital computer modeling, and types of analogous modeling. However, when the proposed mathematical model cannot be verified experimentally, the design of the experiment must be planned much more carefully. Here the experiment would probably be limited to models geometrically similar to the prototype.

In some particular problem areas, where a mathematical model has not been formulated, the use of geometrically scaled models in an experimental program must also involve the critical evaluation and analysis of boundary conditions, environmental conditions, and size effects. This is also true for all other modeling techniques; but for the case where no mathematical model is available, the only check point is one where all extraneous inputs have been either eliminated or evaluated. A boundary condition might contribute to the physical phenomena in the real problem but be eliminated in the modeling. A ship passing through the Panama Canal could be affected by the canal sides, while a scaled model tested in the open ocean would not be affected by the boundary. The environmental conditions of temperature, humidity, pressure, light sources, vibrations, and others have obvious effects on a system, and any geometric modeling can be greatly altered if the effects of the environmental conditions are not also scaled. Finally, geometric scaling can change the physics of the problem by moving to a region where either a phenomenon previously negligible becomes dominant or a phenomenon previously dominant becomes negligible. This is called scale effects. For example, in considering surface-tension effects in a tube of $\frac{1}{10}$ in. inside diameter, geometric scaling might indicate an experiment using a tube $\frac{1}{1000}$ in. inside diameter. However, the act of scaling has possibly changed the physics of the problem from one where gravitational forces dominate to one where capillary forces dominate. When the 6-in. shaft is modeled with a $\frac{1}{2}$-in. shaft, the possibility of buckling exists for compressive loads with a given length-to-diameter ratio. These are only a few examples of the types of extraneous inputs that can occur in an experimental program. If all limitations imposed on experimental results are not known, the information should be used with caution, and a test of the full-scale (prototype) model is indicated.

Several good books on modeling techniques are listed at the end of this chapter. When there is any doubt about the limitations imposed by a given model, these books should be consulted. To develop the subject of modeling beyond what has already been presented would require that the major effort of this book be directed toward this area. Since several references are available, it is not necessary to pursue this topic. However, an attempt will be made to employ a variety of modeling techniques in the examples, particularly in Chapter 4 of this book. This should develop approach techniques for designing the experiments.

2.5 Selecting the Control Parameters

In any experimental program, all variables affecting the experiment should be listed. Then the variables can be divided into classes of those that are to be maintained constant for a particular test and those that may vary. In selecting the parameters to be controlled, it is important to include all variables that significantly affect the test being performed. It is also important to recognize variables that are inflexible in terms of external control. (For example, real time cannot be adjusted; viscosity varies with temperature over a limited range and direct control may be difficult.)

If a proposed mathematical model is to be examined experimentally, control variables are clearly indicated; the relationship between two variables can be obtained by varying the independent variable, maintaining all other variables constant, and measuring the dependent variable. Then, the influence of the other variables can be determined by interchanging the roles of the independent variables. If we were trying to establish the relationship between pressure, density, and temperature for a perfect gas ($p = \rho RT$), we would select an independent variable, perhaps pressure, fix the value of density, and measure the change in pressure with change in temperature. Then we could fix pressure and select density as the independent variable to obtain the relationship between temperature and density at constant pressure. We could actually continue using the original parameters selected (pressure independent and temperature dependent). Then we could determine the relationship between these variables for several different but constant values of density. Hopefully, there has been some mathematical model proposed for parts of the problem. However, a frequent situation occurring in an experimental program is that where no mathematical model has been proposed. The method of selecting control parameters is no longer simple, but the technique of using a dimensional-analysis approach has frequently been profitable.

The subject of dimensional analysis deserves much more complete coverage than is possible without changing the basic intent of this book. Therefore, the readers are directed to the references at the end of this chapter for detailed coverage of the subject. A very brief presentation of dimensional analysis with some examples is also presented in Appendix II.

It would be informative to demonstrate the four steps involved in the design of an experiment. To be more meaningful, the same problem will be solved for a progressive variety of accumulated technical knowledge.

EXAMPLE 2.1

A 6-in.-diameter steel shaft is to be used to support a cornerstone of a building. Will it be adequate for 200 years' usage? The building will require 10 tons support on this corner.

1. Defining the problem—if the problem is not clear in every detail, now is the time to establish reasonable bounds. This problem is reasonably complete, but the environmental conditions might require some elaboration. When incomplete information is available at this point, either time must be spent in collecting the desired information, or an assumption must be

made regarding the unknown items. An engineer with no previous experience in an area would undoubtedly do a thorough job of determining the unknown information. However, once experience has been gained, much of the unknown information can be approximated with an accuracy adequate for the problem solution. Therefore, the assumptions made in this example will be used for this particular example with confidence that they are correct. Production costs apparently do not enter the problem, since the product decision has already been made. It is reasonable to assume that the chemical environment is inert. Temperature and wind conditions could change the load (however, for the present problem the coefficient of thermal expansion and the building's aerodynamic profile will be neglected). One other problem that should be considered is the vibration that might result from either automobiles operating near the building or natural phenomena (earthquakes).

With these limitations in mind, it is possible to search the literature. For the present problem, this search will result in three different degrees of success: a different result for each of examples 2.1, 2.2, and 2.3.

2. Assimilation of technical knowledge—Part I—The literature review resulted only in the knowledge that material failed when too much force was applied (of course, this literature review was taken before the time of Hooke).

3. The design of the experiment would probably involve either model or full-scale testing. The design would include boundary (loading) conditions, environmental conditions, and size effects (if a scaled model is used). It is important to know the interfacial loadings between the cornerstone and the steel. This problem might also require an experimental solution, but for the present it will be assumed that the loading in the experiment is identical to that of the building.

All of these problems must be considered in a real experimental program. When everything is not known, however, the tendency is to overdesign rather than spend a tremendous amount of time discovering that an effect is negligible. For the technical knowledge given in this part, it is obvious that essentially nothing is known about this problem. For this case a dimensional-analysis-type approach is suggested.

4. The variables affecting the steel shaft are:

1. Load, F
2. Diameter, L
3. Length, L
4. Material, FL^{-2} (modulus of elasticity)
5. Change in length, L

The Buckingham π theorem determines that there are three dimensionless groups that affect this problem (number of variables, 5, minus number of primary dimensions, 2). Of the groups possible, the three that will be used are

$$
\begin{array}{ll}
\text{Change in length/length} & L/L \\
\text{Length/diameter} & L/L \\
\text{Load/area/material} & \dfrac{F/L^2}{F/L^2}
\end{array}
$$

Remember that selecting the control variables involves both a list of the important variables and consideration of variables that are easily controlled. Length and loads are easily controlled and easily measured. These are, therefore, selected to play the role of the control variables.

Since only one length/diameter is considered in this problem, that dimensionless ratio is constant. (This dimensionless group becomes important if a buckling condition exists.) Then a change of the loading could be used to obtain the relationship

$$
\frac{\text{change in length}}{\text{length}} = \frac{\text{load/area}}{\text{material}}
$$

or

$$
\epsilon = \frac{S}{E}
$$

Several comments are in order to clarify the status of the example problem at this point.

1. Vibration effects were not included in the list of variables; therefore, this is a static problem.
2. We could easily have solved the vibration problem to obtain the results of Example 2.3, which follows this example. However, it was desirable to keep the problem simple here.
3. The selection of the dimensionless groups was based on some experience and should not be interpreted to be a straightforward reduction from the Buckingham π theorem. (For example, load/(length)²/material is also obtained by applying this theorem.)
4. The concept of buckling could not be obtained from the present experimental program.
5. The effect of life (200 years) has not been considered.

EXAMPLE 2.2

Assume the same problem statement (Step 1) as Example 2.1 with the same defining characteristics. However, suppose the literature review (Step 2) gives the fact that

$$S = \epsilon E$$

Now, what is the experimental design and what are the control variables, assuming that the modulus of elasticity, E, is known for the steel of your problem?

3. It is still important to test your specific case, probably using a geometrically scaled model. However, an analogous system could be used. For example, a constant-voltage battery (E) connected to a variable resistor $(1/\epsilon)$ could be used to control the current (S) flowing in the circuit, $V = IR$ or $E = (S)(1/\epsilon)$. For this particular system the scaled model would probably be used. A test is still necessary, because all limitations of the equation $S = \epsilon E$ were not obtained from the literature review; and it is possible (although not for this particular problem) that failure could occur by buckling.

If vibration effects are still neglected, the experimental design would involve a geometrically scaled model (possible $\frac{1}{12}$ to $\frac{1}{24}$ of full scale). The control variables (Step 4) would be the same as those of Example 2.1, but the test would require fewer points since the mathematical law has already been established. This is a much simpler solution than that of Example 2.1, since it requires no personal experience to find the proper dimensional groups.

Again some comments might be valuable.

1. Neither solution considered the problem of creep (if the structure is to stand for 200 years, creep is a possibility).
2. Both solutions neglected the fatigue loading due to vibrations.
3. It is not extremely obvious that Example 2.2 is much simpler than Example 2.1, but several attempts to select the proper dimensionless groups from a dimensional-analysis approach will help to make this clear.
4. The experience suggested in comment 3 will be sufficient motivation to ensure that a complete literature review be made before the experiment is designed.

EXAMPLE 2.3

For the same problem (Step 1) of the previous examples, suppose that the literature review (Step 2) reveals the relationship

$$M\ddot{X} \times C\dot{X} + KX = F(t) \tag{2.3}$$

3. This is the differential equation governing the motion of the steel shaft when the input loading or stress varies with time. When the basic differential equation is known for a phenomenon, the experimental problem simply becomes a matter of verifying the validity of the equation and establishing the limits of boundary conditions, environment, and size effects. Many methods of approach are available once these limits have been determined. The equation is analogous to the mechanical vibration system and the electrical circuit system of Equation 2.2.

If Equation 2.3 governs the motion of the steel shaft, either a digital computer or an analog computer could be used to predict the behavior of the steel shaft under various loading conditions. Considerably more could be determined about the problem in a very short time using either of these methods. Actually, if the governing differential equation is known along with its limitations, the problem is no longer an experimental problem. The full-size solution would probably be tested in operation if human life were not endangered.

However, for the present problem the constants in the equation are not easily obtained, and the problems of creep and fatigue have not been resolved. Therefore, the ideal solution could be obtained by a literature review which included these two phenomena; or they must be determined by performing tests on geometrically scaled models.

It is interesting to show how the differential equation can be used to obtain the dimensionless groups useful in similarity modeling and for dimensional analysis. In this case Equation 2.3 is used. The governing equation is

$$MX'' + CX' + KX = F(t)$$

where $X = x - x_0$ and x_0 is the length of the unstressed member. Now, K is seen to be equal to EA/x_0, where A is the cross-sectional area. This equation is nondimensionalized by letting

$$\bar{X} = \frac{X}{X_0} \quad \text{and} \quad \bar{t} = \frac{t}{t_0} \qquad \text{(where } X_0 = x_0)$$

The equation becomes

$$\frac{MX_0}{t_0^2}\frac{d^2\bar{X}}{d\bar{t}^2} + \frac{CX_0}{t_0}\frac{d\bar{X}}{d\bar{t}} + KX_0\bar{X} = F[t_0(\bar{t})]$$

For a particular loading function $F(t)$, the function can be made to depend on the dimensionless time, $\bar{t}$, only. Then, if both sides are divided by the shaft area, the equation becomes

$$\frac{MX_0}{At_0^2}\bar{X}'' + \frac{CX_0}{t_0 A}\bar{X}' + \frac{KX_0}{A}\bar{X} = S(\bar{t})$$

From the static case $KX_0/A = E$. Since all terms $\bar{X}, \bar{X}',$ and $\bar{X}''$ are dimensionless, division by the coefficient of any one of them will give a dimensionless equation. Suppose both sides of the equation are divided by E to give

$$\frac{MX_0}{At_0^2 E}\bar{X}'' + \frac{CX_0}{t_0 AE}\bar{X}' + \bar{X} = \frac{S}{E} \tag{2.4}$$

Then, the dimensionless groups of importance are the $\bar{X}$'s, the coefficients of the $\bar{X}$'s, and S/E. If the differential equation is known, this method always produces the dimensionless groups necessary for similarity modeling. However, this method does not consider the effects of boundary conditions, environment, or size effect where the couple forces cannot be neglected if the diameter is much smaller than the length in this problem.

4. The control parameters are again the same as those selected in Step 4 of Example 2.1.

For Equation 2.4, similarity modeling between a model and the prototype can be obtained by requiring equality of the dimensionless ratios

$$\left(\frac{MX_0}{At_0^2 E}\right)_{prototype} = \left(\frac{MX_0}{At_0^2 E}\right)_{model}$$

$$\left(\frac{CX_0}{At_0 E}\right)_{prototype} = \left(\frac{CX_0}{At_0 E}\right)_{model}$$

$$\bar{X} = \left(\frac{x - x_0}{x_0}\right)_{prototype} = \left(\frac{x - x_0}{x_0}\right)_{model} \tag{2.5}$$

$$\bar{X}'_{prototype} = \bar{X}'_{model}$$

$$\bar{X}''_{prototype} = \bar{X}''_{model}$$

and

$$\left(\frac{S}{E}\right)_{prototype} = \left(\frac{S}{E}\right)_{model}$$

For the static case using the same material for model and prototype (i.e., $E_p = E_m$), the stress in the model must equal the stress in the prototype and the strain, $\bar{X}$, in the model must equal the strain in the prototype. By using different materials, it is possible to predict what would happen with the steel shaft by using a type of plastic. In any situation, Equations 2.5 must all be satisfied for similarity to be applicable.

2.6
Execution of
the
Experiment

After all of the planning of the experimental program has been completed, the measurement systems must be selected according to the information in Chapter 3. Then the experiment can be performed and the resulting data can be evaluated. During the actual performance of the experiment, new knowledge can direct the course of the experimental program. The amount of experimental data required in a given range of parameter values can often be best decided during the actual data-collection period. The overall success of experimental analysis is a result of each individual element in the program.

During the execution stage of the analysis, it is possible to begin an evaluation of the program. Frequently, this information can be put to use immediately to establish areas of strength and weakness. Additional attention can be directed toward the weakness at this point in the analysis. Example 2.3 represents the power obtained from similarity laws when the actual execution begins.

A knowledge of the laws of similarity can greatly reduce the amount of data necessary to establish a relationship valid for a variety of situations. For static loading the equation $S = \epsilon E$ has proved adequate for representing a restricted range of stress–strain information. This equation can be made dimensionless by dividing both sides by the modulus of elasticity to obtain $\epsilon = S/E$. It is not necessary to test specimens of various diameters in tension to obtain a curve valid for all diameters. Therefore, a test of 10 different diameters would multiply by 10 the amount of experimental time that would have resulted in the same experimental verification. It should be pointed out, however, that a test of two or three diameters might be called for if the tests were for compressive loading, where the possibility of buckling exists.

The use of similarity principles can be a powerful tool in saving time. When there are some doubts concerning the applicability of similarity laws to a given situation, it will be necessary to expand the experimental program to include enough data to evaluate the accuracy of using similarity. This limitation must be considered again in the section on analysis of the experimental program in Chapter 4.

2.7
Conclusions A detailed study of design of the experimental program could not be included in this book within the space limitation. Two methods for developing strength in this area of experimental analysis are available:

1. The experimental examples used in this book will attempt to elaborate on the proper techniques for designing the experiment.
2. The references listed at the end of this chapter provide much more complete information for outside study.

The problem areas you may experience in the design of an experiment can be discovered by following the simple outline given in this chapter. A block diagram of this procedure is provided below for the convenience of the user.

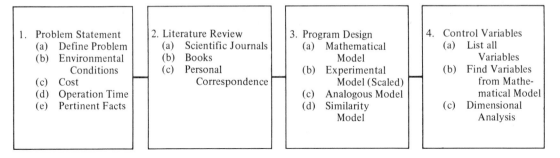

Figure 2.1 Block Diagram of Experiment Design

References

1. BARTEE, E. M., *Engineering Experimental Design Fundamentals*, Prentice-Hall, Inc., Englewood Cliffs, N. J., 1968.

2. Cos, D. R., *Planning of Experiments*, John Wiley & Sons, Inc., New York, 1958.

3. IPSEN, D. C., *Units, Dimensions and Dimensionless Numbers*, McGraw-Hill Book Co., Inc., New York, 1960.

4. KEMPTHORP, O., *The Design and Analysis of Experiments*, John Wiley & Sons, Inc., New York, 1952.

5. LANGHAAN, H. L., *Dimensional Analysis and Theory of Models*, John Wiley & Sons, Inc., New York, 1951.

6. SCHENCK, H., JR., *Theories of Engineering Experimentation*, McGraw-Hill Book Co., Inc., New York, 1961, 1968.

PROBLEMS

2.1.   Perform a library research to obtain a list of publications for one of the following topics, or select some additional topic of interest and find the work done in that area.
(a) Machine tool–work piece thermocouples
(b) Transient pressure calibration
(c) Nucleate boiling heat-transfer correlations
(d) The measurement of surface roughness
(e) Experimental stress concentration
(f) Space simulators
(g) Thermal modeling
(h) Aerodynamic modeling

2.2.   Determine all the independent dimensionless groups that can be formed from the following physical variables: density ($P = ML/\theta^3 T$), viscosity ($\mu = M/L^3$), characteristic length (L), velocity ($V = L/\theta$), thermal conductivity ($K = ML/\theta^3 T$), specific heat ($C_p = L^2/\theta^2 T$), heat-transfer coefficient ($h_c = M/\theta^3 T$).

   Use as the primary dimensions:

$$
\begin{aligned}
\text{mass} &\longrightarrow M \\
\text{length} &\longrightarrow L \\
\text{time} &\longrightarrow \theta \\
\text{temperature} &\longrightarrow T
\end{aligned}
$$

2.3.   In the flow of fluid in a tube, the significant variables are pressure drop, pipe length, fluid density, fluid velocity, fluild viscosity, tube diameter, and roughness of the tube surface. What dimensionless groups should be important in this problem?

2.4.   A one-tenth scale-model car is tested in a wind tunnel with air at standard conditions: (a) If the viscous drag on the model tested is 0.2 $1b_f$, determine the actual drag on the prototype traveling under the same conditions. (b) If the air velocity in the wind tunnel is one-half that of the prototype, determine the drag on the prototype when

measurement system. Measured values are meaningful only when they are properly collected and evaluated. Therefore, the entire body of knowledge is very closely related in an extremely complex manner. The logical method of presenting this material is to develop each portion at an equal rate, but two disadvantages of this type of presentation are that it repeats previous work and it is slow. In view of the constraints, the subject of design of the measurement system will be developed at the expense of the equally important design of the experiment and analysis of the program.

In attempting to present a complete and detailed discussion of the measurement system, the magnitude of the portion of the book devoted to this subject tends to distort the true perspective of its relative position in experimental analysis. Remember that the only purpose of the measurement system is to obtain reliable values for the variables or parameters of the experiment being considered. The word reliable necessitates a complete analysis of the measurement system. The function of the measurement system is shown in block-diagram form in Figure 3-1. All three parts of

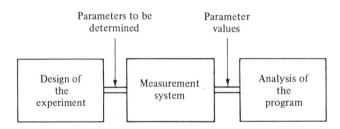

Figure 3.1 Functional Components of Experimental Analysis

experimental analysis are equally important in resolving the basic problem. However, this book has been directed principally toward the problems involved in obtaining reliable values for the parameters being considered. This chapter will cover the basic concepts of the measurement system. Chapter 4 will develop a problem-solving format. Chapter 5 will deal with program analysis, and then a more detailed look at measurement systems will be considered in Chapters 6–9.

3.2
The Significance of Measurement

Measurement is defined as the comparison of an unknown quantity with a known standard. A measurement system accomplishes the act of comparison. The standards used in this comparison process will be completely defined later in this chapter. However, standards are the established, recognized, accepted, and invariant quantities used in a comparative process called calibration.

A complete measurement technology must include standards for the measurement of any parameter that could be encountered. From our basic dimensional system, this would require standards for length, time, mass, and probably temperature in the minimum case. Force could be determined from Newton's second law, and electrical quantities could be determined from the basic definitions of these quantities. Practically, the difficulties involved in using these definitions for electrical quantities have led to secondary standards for current and voltage.

One very significant fact to bear always in mind is the first axiom of measurement. This axiom states that the very act of measurement changes the system being measured. For example, placing a thermometer in a room to determine the temperature of the room changes the room conditions in two ways. First, some air must either be pushed out of the room by the volume occupied by the thermometer, or the room pressure increases. Second, the thermometer will not be in absolute thermal equilibrium with the room when it is first placed there and will consequently change the temperature of the room by its presence. Many systems can apparently violate the first axiom of measurement because their effect on the system is extremely small, and these are the types of systems sought for use. A light-wave pattern might be used to determine the strain in a solid. However, when the light strikes the surface being considered, there is an energy exchange that changes the surface.

The discussions that could arise over the validity of the first axiom of measurement can be heated. That is not the purpose of introducing the axiom here; the purpose is to make sure the reader is aware that his attempt to measure a parameter almost always introduces a significant change in that parameter. At least he should always consider this possibility.

Historically, the advancement of measurement technology has paralleled that of scientific progress. The accomplishments of the Egyptians were a result of great architecture and the measurement of length. As measurement became more precise, the limitations of the mathematical models of physical laws were discovered. These discoveries usually lead to a formulation of more complete models. Throughout history, the capabilities for making measurements have always significantly affected the living standards. Commerce depends to a great extent on the measurement standards. Trade between countries would be controversial if present standards did not exist. Daily purchases at the local market are made without price bargaining because both parties know the value of the product.

Daily life is measured out with relatively precise time-measuring devices.

The busy schedule of today requires careful planning of each day's activities. The problem of overweight people is monitored (using bathroom scales) to indicate when some additional controlling action might be necessary. Swatting a fly, cooking by recipe, throwing or hitting a ball, feeling the temperature or humidity of the air, planting landscape, and building anything are all daily examples of measurement. With advances in the technological field, daily living requires more and more measurement capabilities. Measurement has made a significant entry into the daily lives of the nation along with the tremendous requirements of a highly advanced scientific technology.

3.3
Elements of a
Measurement
System

It is always important to keep the measurement system as simple as possible while still accomplishing the required job. Therefore, measuring a refrigerator to determine whether or not it will fit a given space could be accomplished using a yard stick, while adjusting the point gap setting of an automobile ignition system could not be accomplished using this crude measuring device.

All measurement systems have similar functional components. The usual method of functionally dividing up the measurement system is to consider the purpose of any measurement element. The system is to determine the values of various parameters. This requires at least two functional parts: one part to sense the parameter to be measured, and the second part to present this sensed information in a meaningful form. To obtain the information in a meaningful form may require some modification. The measurement system may be represented by the block diagram shown in Figure 3-2. Each of these components will be considered in detail before additional material is introduced.

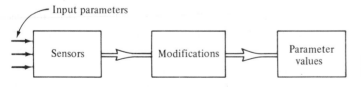

Figure 3.2 Elements of a Measurement System

To obtain a reliable value for desired parameters using the type of system shown in Figure 3-2, there must be a standard used at some point in the system design, a complete calibration procedure must have been performed, and a complete analysis of the data taken must be available. To simplify the analysis, the measurement system will be considered for each input parameter separately. Since the design of the overall measurement system is completed by designing a system to measure each of the significant parameters, the analysis of a system to measure a single input parameter is completely general.

In discussing the elements of the general measurement system of Figure 3-2, it is quite natural to start with the sensor which is the first element actually to be involved with the measurement process. This procedure will be followed here; however, to observe the performance of the sensors discussed, it will be necessary for the reader to understand the operation of several types of read-out devices. Therefore, it may be convenient to skip over to the operation of such simple output devices as the oscilloscope, voltmeter, voltage-balancing potentiometer, electronic counter, and scale–pointer devices at this time. This will depend on the background of the reader and the laboratory experience that is associated with the reading of this book.

Sensing Devices Perhaps the most elaborate, versatile, and reliable measurement system available is the human being. His five senses immediately detect a wide variety of input parameters and modify these sensed signals to a form that is readily available for use. The eyes can sense distance, angle, vibrations (with light reflecting off of the vibrating object), colors, convective currents, humidity (when 100 per cent), and a multitude of other parameters. The accuracy of these sensed parameters and the corresponding output values vary primarily with the experience of the human in sensing these signals. The experienced operator of a blast furnace could often tell when molten metal was ready to be poured better than the first measurement systems devised. The experienced tire-balancing expert can sense dynamic unbalance by observing the vibrations of light off the bumper while the wheel is jacked up off the floor and driven at high rotating speeds. He may also sense this signal by the touch of his hand. Both methods can be extremely precise.

The human measuring system is like all measurement systems in that it is very easily deceived by physical phenomena. For example, hold a pen 2 ft from your face and alternately open one eye and close the other. The pen appears to move from side to side. Try to judge the velocity of a bird or airplane. The old trick of saying "look out for that hot coal" and dropping a cube of ice on a person is an example of deceiving the human measurement system. Any measurement system can have a certain number of liabilities or extraneous inputs. There are two very important reasons why it is essential that the functional operation of the sensor be completely understood. First, if all considerations necessary to establish an accurate value for a parameter are to be made, the response of the sensor to the total environment must be known. Second, every element of the measurement system can be seen (with very little insight) to be functionally identical to the operation of a sensor. Therefore, every consideration and discussion concerning the performance and operation of the sensor can also be applied directly to the other two elements of Figure 3-2.

To sense or detect a given physical parameter or quantity, a sensor must respond to this parameter in some form. If the sensor is to be useful for measurement, it must respond in a prescirbed manner to a given input parameter. In mathematical terms, this means that there exists a relationship which connects some measurable parameter of the sensor to the parameter to be measured. In word equation form,

$$\text{sensor output parameter} = \text{function of parameter input information}$$

or

$$\text{information out} = F \text{ (information in)}$$

The fact that a functional relationship exists between input and output information is sufficient to ensure that the sensor is capable of detecting the desired parameter. Measurement-system calibration is the process of establishing the functional relationship that exists between input and output information.

The selection of a sensing element is based on six major requirements determined from a knowledge of the problem being solved.

1. The sensor must be sensitive to the variable to be measured.
2. The sensor should be insensitive to all other inputs.
3. The accuracy and range requirement of the measurement must be known.
4. The dynamic response of the sensor is required.
5. The cost of the entire measurement system is to be considered.
6. The complexity of the modifying and recording elements required for each type of sensor must be considered.

Each of these six requirements must be satisfied by the sensor that is ultimately selected for use. It is helpful to consider each of these requirements in much more detail to provide a complete understanding of sensor elements.

To obtain a sensor that is sensitive to the variable to be measured, every physical law that has some form of dependence on that variable should be listed. One example of this technique follows.

EXAMPLE 3.1

The variable to be measured is force. A partial list of the physical laws which involve force is

1. Newton's law $\qquad F = \dfrac{d(mv)}{dt}$

2. Spring model $\qquad F = KX$
3. Pressure definition $\qquad F = PA$
4. Stress definition $\qquad F = SA$

5. Gravitational attraction $\qquad F = \dfrac{km_1 m_2}{r_2}$

6. Laws of statics $\qquad \sum F = 0, \ \sum M = 0$

7. Torque definition $\qquad F = \dfrac{T}{r}$

There are many other physical laws which have force as a variable in the model, but the present list is sufficiently complete to provide a wide variety of force-sensing elements. Now it is necessary to find a real physical system that is governed by one of the above laws. Any system satisfying this condition could be used to sense force. Essentially, every mass obeys Newton's law for speeds much less than the speed of light. Therefore, any mass could be used to sense a force.

In view of the almost limitless extent of possible physical systems capable of sensing force, an arbitrary reduction was made to consider a variety of sensors having different properties with respect to the other five requirements of a sensing element. This reduction is not intended to imply that the systems not considered here are any less useful than the ones considered. New and amazing sensors are developed each day based on the laws just discussed, and these rapid developments make understanding the fundamentals of the operation of sensors important to the selection of a measurement system.

The systems selected to be considered in this example are shown in Figure 3-3. In part (a) the acceleration of the cup in the direction shown results in a displacement of the liquid level, and for steady (constant) acceleration the angle of inclination determines the acceleration. This phenomenon is governed by Newton's laws of motion (first law listed in this

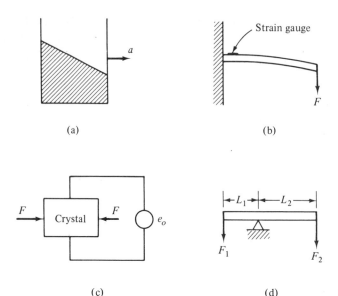

(a)

(b)

(c)

(d)

Figure 3.3 Physical Systems for Sensing Force

example). If the entire mass of the cup and water were known, the force could be determined. In part (b) a force in the direction shown would cause a bending of the cantilever beam. The actual motion of the end of the beam is governed by the spring model, number 2 on the list of laws. However, the motion of the end of the beam causes a stress to occur along the beam, and this is sensed by the strain gage at its location. Part (b), therefore, uses law 2 and one not listed, $S = Mc/I = FLc/I = \epsilon E$. The measurement of strain, beam length, modulus of elasticity, beam width, and beam depth could be used to determine force. In part (c) law 4 is used to relate force to stress and area. The piezoelectric crystal produces a static charge that is proportional to the strain in the crystal. Part (d) uses law 6. This might also be called Newton's first law. It provides a method for comparing two forces. If one force is known, the comparison can measure the resultant of the unknown force. If the moments and forces are applied to a beam balance, one fundamental force-measuring device results.

The above examples demonstrate methods available for considering systems that could be used for the sensing of a desired input signal. It is definitely not necessary to know the exact formulas relating the quantities that can be measured to the parameter of interest. If a functional relationship exists, it can be evaluated experimentally. However, it is extremely helpful to know the exact relationship when the question of sensitivity to other inputs must be answered.

For a system even to be considered for use to measure a given parameter, it must satisfy requirement 1; that is, it must respond to the desired parameter. Ideally, the system is insensitive to all other inputs, but this ideal is seldom attained.

One of the most complex problems to be solved by the measurement-system designer is the elimination of extraneous inputs to the sensor system. The first and most logical starting point for reducing extraneous inputs is to use sensors with this capability inherent. Each physical system will now be considered with respect to its sensitivity to other input signals.

EXAMPLE 3.2

Sensitivity to Other Inputs. Water in a cup is sensitive to accelerations in any direction, but it is doubtful that a meaningful measurement could be made if transient accelerations were present (this is also evidenced by the spilled coffee spots around the laboratory). If the temperature is greater than saturation temperature, boiling would create an extraneous input. It is also sensitive to the viscosity of the water and other fluid properties (surface tension, temperature).

The strain gage on the cantilever beam is sensitive to all of the inputs that affect the beam. The modulus of elasticity and area change with tempearture, permanent set, creep, fatigue, and yielding. Any of these inputs would change the output from the desired signal. Additionally, the strain gage is subject to all of the above types of extraneous inputs; and the final output is a complex composite of the accumulated inputs. The strain gage is also sensitive to the current flow required to measure strain-gage resistance. Any electrical noise, electrical or magnetic fields, and thermoelectric effect could also cause extraneous inputs to the strain gage.

The piezoelectric crystal is sensitive to electrical noise, electrical or magnetic fields, lead line–crystal interfacial condition, and temperature. The crystal also allows the electrical charge to leak (or move back to its original position) if the input force is a constant. The crystal is sensitive to pressure waves, vibrations, and any accelerations of the crystal.

The beam balance is sensitive to the variation of knife-edge dimensions with both temperature, wear, and any material properties of the beams. It is also sensitive to bending of the beams, vibrations, magnetic fields, yield, and permanent set.

The lists of inputs to which each of the above sensors might respond is not complete, but it should give insight into the type of problems that must be recognized in this field. Every extraneous input listed for each of the examples is a variable that could be detected by the sensor described. It is desirable to have a system which is completely insensitive to all inputs except the input of the variable to be measured. If a system is sensitive to other inputs, it is necessary to analyze in detail the variation in the parameter being measured that results from an extraneous input. In some cases the problem is so complex that the only method that approaches a satisfactory solution is a completely experimental verification of the effect. When the sensor system is so sensitive to extraneous inputs that the variable to be measured is completely concealed by the extraneous signal, the sensor is unsatisfactory to measure the desired variable; but it might be a useful sensor to measure the extraneous signal.

The magnitude of the interfering or extraneous input determines the feasibility of using a particular measurement sensor to detect a particular signal. For the systems considered above, more information concerning the conditions under which the measurement system is to function must be available, along with more information on the magnitude of the interference, before a complete evaluation can be made. A typical criterion for selecting a measurement system is to require a signal-to-noise ratio of approximately 10 to 1 for satisfactory operation. Here the signal is the output resulting from a change in the variable to be measured, and noise is the extraneous input.

In considering the accuracy and range in which a given sensor is capable of acceptable performance, preliminary consideration of the capabilities of the complete measurement must be made. A pure signal (one that is insensitive to all normal extraneous inputs) can frequently be magnified by the other elements of the measurement system. A very small pure signal is usually much more valuable than a large signal composed of the desired signal and a multitude of extraneous signals. The techniques of filtering and compensating for extraneous signals can often be employed to produce a signal that performs as accurately as a pure signal.

The accuracy of a sensor is actually very complex and can be defined in many terms. From the viewpoint of an individual designing a measurement system, the primary concern is the number of significant figures obtainable in a measured value of a variable. Accuracy for him, then, is the number of significant figures that can be derived from the modified signal, which may have been filtered or may have involved some compensation procedure. A number can be significant only when it is repeatable for the same input signal. Therefore, the type of input signal will also make some contribution to the value assigned to describe the accuracy. It might be necessary to determine an accuracy figure for a static signal input, for a sinusoidal input signal, for a step-change input signal, and for a reproducible transient input signal. These values for each type of input may be different. The more complete evaluation of accuracy will be considered later in Chapter 5. This present discussion should present a feel for the factors that affect accuracy.

The range of values of inputs over which the sensor is capable of operating depends on three principal factors. First, over what range of inputs is the sensor capable of detecting the desired input? Second, over what range of inputs can the sensor provide a signal of sufficient magnitude and purity to be useful for the measurement? Finally, over what range of inputs can the sensor signal provide a signal that can be modified to result in the desired accuracy? Answering these three questions also requires some insight into the other elements of the measurement system. Again, the limitations may be confined to particular types of inputs. Therefore, it is also necessary to specify the range limitations by each type of input signal. Normally, the range will be specified according to an analysis of the physical limitations such as temperature, strength, elastic limit, and permanent sets. These limitations will also be considered in very general terms for the systems of the above example.

EXAMPLE 3.3

Accuracy and Range. The accuracy of the water–cup sensor is dependent on the accuracy with which the angle of inclination of the water can be measured. It is also extremely dependent on the condition of a constant acceleration. Any fluctuations in input acceleration make the angle of inclination unreadable. The range of operation is limited by the fact that the water will spill out if the acceleration is too large. A lid on the cup could eliminate this problem. Another range limitation is the requirement that the velocity be small compared to the speed of light. (This may be an unrealistic restriction, but Newton's law must be modified at this velocity, and this is certainly an upper limit that should be recognized.)

For the cantilever beam, the accuracy depends on the sensitivity of the spring (cantilever beam), which will change with temperature. The accuracy is determined by the relative displacement of the beam and the accuracy of measuring strain. The type of input signal will produce various degrees of accuracies. The range is limited by the elastic limit of the cantilever beam, the temperature of the beam, the bond between the strain gage and the beam, property limits on the gage material, and lead-wire connections. The range of input signals is also affected by the natural frequency of the beam.

The accuracy of the piezoelectric crystal depends more on the other elements of the measurement system than on the sensor. Accuracy can be affected by temperature. The range of use is generally limited to the measurement of dynamic (or rapidly changing) forces. There are also fixed temperature ranges for any given crystal. The range is also limited by the smallest force detectable and by the elastic limit of the crystal.

For the beam balance, the accuracy of measuring one force and the two lengths involved limit the overall accuracy. The physical observation of balance is subject to the accuracy of the observer. The dimension of the knife edge is important. The range is limited to static inputs of magnitude small enough to avoid bending and failure. Temperature limitations are the same as for all devices involving metal parts (melting, brittleness), with the added problem of changing the deformation of the knife edge.

In general, a change in temperature, humidity, or pressure will cause some change in the accuracy of a measurement. Accuracy is often limited by the scale divisions available on an output device. The number of significant figures that can be read from a slide rule is analogous to the number of significant figures that can be read from a pointer-dial output device. The accuracy can never be better than the number of significant figures obtainable from an output device. Range is generally limited between the smallest signal detectable (the threshold signal) and the signal so large that it overwhelms the system. If an output for a particular measurement system is zero, this does not mean that the parameter value is zero. It simply means that the value is less than the threshold signal for this system. Temperature, inertia, and friction can also play dominant roles in the range over which a system can function. Most of these limitations are obvious, but many systems are designed utilizing elements that are obviously incapable of performing over the desired range.

Accuracy and range in all of the above cases always depends on the type of input signals. Static signals are the easiest to measure and evaluate. All dynamic or transient input signals are affected to some degree by inertia. The greatest problem in determining the accuracy and range of a system is the problem encountered with a changing input signal. This problem is considered in Appendix III on calibration. A simple consideration of the dynamic response of the example systems will be included in this section.

Any measurement system which exactly reproduces the dynamic input is called a zeroth-order system. This system is the goal or ideal system. Once this goal has been realized in a measurement system, the problems of measurement and data analysis have been resolved in general. The few remaining problems can be solved by applying much simpler analytical techniques. It should be realized that all real measurement systems cannot exactly reproduce the input signal at the output without distortion of some kind.

The types of distortion are normally considered in terms of wave technology. In this terminology the categories are phase distortion, amplitude distortion, and frequency distortion. The examples of force sensors will be analyzed to broaden the concepts of these distortions. While a perfect (zeroth-order) system will have none of these distortions, the real system will distort the input signal with varying degrees in each of these three classifications. Many systems are quite capable of measuring static signals, but they are useless in the measurement of dynamic signals. The factors that control the type of input signals that can be measured must be recognized if they are to be applied to any different systems.

Dynamic-signal measurement is one of the most challenging problems in measurement. A complete analysis of the system, including the basic governing dynamic equations, in some cases allows the use of distorted signals for making valuable measurements. However, sometimes the response of a system is so sluggish that no useful information can be obtained from the measurement. For example, a mercury manometer cannot be used to measure the pressure change across a moving shock wave. The human thumb is not fast enough to use a stop watch to determine the time required for a cycle of motion in an automobile engine. A feather cannot be dropped in air to determine the acceleration due to gravity. More complete examples are considered for the systems that have already been explained.

EXAMPLE 3.4

Dynamic Response. The cup of water is basically a static-acceleration measuring device. If the acceleration is either increasing or decreasing at a constant very slow rate, it might be possible for the water level to respond sufficiently to obtain an accurrate measure of the acceleration. When the acceleration increases and then decreases rapidly, or when the acceleration changes sign rapidly, the water tends to slosh, and no plane can be established. Ripples in the water indicate a change in the acceleration, but the value of this change cannot be determined.

The dynamic response of the cantilever beam requires a much more careful analysis. This result also applies to all spring–mass systems. The student is familiar with the dynamic equation governing a spring–mass system with no damping but having a forcing function

$$m\ddot{x} + KX = F(t)$$

Recall that the solution to this problem is composed of a complementary solution and a particular solution. A complete solution and analysis is presented in the calibration section, Appendix III, but with zero damping the complementary solution does not vanish. Therefore, the distortion introduced by the complementary solution continues to disrupt and interfere with the input forcing function to be measured. This causes phase, amplitude, and frequency distortion. Because the mathematical model of this system is well known, an evaluation of the output could provide adequate measurement information for constant acceleration and impulse-type acceleration. However, the analysis becomes quite involved. In real systems some damping is always present, and the complementary solution will essentially vanish with time. This added factor complicates the analysis for rapidly changing input, but it can prove extremely useful when the damping takes on certain values. Most measurement systems contain a spring–mass system somewhere in their elements. For this reason an understanding of the fundamentals of this system will be used throughout the balance of this field of study. In terms of dynamic response, this system is called a second-order device. Note that the governing equation is a second-order equation, and the reason for the response classification becomes clear.

Both the piezoelectric crystal and the strain gage placed on the cantilever beam have the same limitations to dynamic response. These systems will always produce an output signal that lags the input signal by the time required for the input force to cause the elongation. This is essentially the time required for a sound wave to travel through the metal or crystal (approximately 18,000 ft/s). With a $\frac{1}{4}$-in. piezoelectric crystal and with a strain gage length of $\frac{1}{4}$ in., the time required to respond to any type of input signal is approximately 10^{-6} s. The many advantages and operational capabilities of the strain gage and the piezoelectric crystal will be considered in Appendix III. Both devices are reasonably close approximations to zeroth-order system for ranges of input signals.

The beam balance is essentially capable only of measuring static loads. If the friction of the knife edge is considered, the system is a second-order system with damping, but the motion of the pointer makes visual reading extremely difficult. The only way that a pointer and dial can be used to measure dynamic signals is to use an additional type of read-out device. It might also be helpful for the reader to think of different measurement systems capable of performing the required measurement of force. There are many methods available, and each can challenge the mental gymnastics of the reader to more rapid advancement. This technique should not be limited to the single example considered above. At this point the reader is capable of applying the techniques to each parameter to be measured. Active participation will result in a deeper appreciation of the problems facing the measurement engineer daily. Since it is impossible to present every possibility in this text, the development of an independent approach to the measurement problem can be extremely advantageous in the resolution of future problems.

Eventually, under a discussion of dynamic response, the advantages possessed by electrical measurement systems are recognized. The basic hindrances to rapid dynamic response are inertia and friction. The mass of the electron and the relatively void space it sees reduce the magnitude of both of these difficulties to a minimum. The same considerations have also led to the increased utilization of measurement systems having photons of light, wave characterstics of light, and electromagnetic waves as their elements. In addition to dynamic-response advantages, this type of system is theoretically capable of more accuracy and less disruption of the system being measured. The electron-beam oscilloscope is one of the most often used read-out devices, because it operates using the mass of the electron.

Many measurement systems are required to measure only static signals. For these cases it is not necessary to have elaborate systems capable of measuring rapidly changing dynamic signals. The static system reduces the complexity of the measurement considerably. However, the introduction of dynamic measurements requires consideration of dynamic response. The primary factors affecting the measurement of a dynamic signal are inertia and friction (damping). Each analysis of a dynamic system must take cognizance of the significant roles played by these parameters in the performance of the system.

The next consideration in the selection of a sensor is the probable cost that this sensor will impose on the total system. Each sensor requires characteristic modifying and recording systems. Therefore, the selection of a sensor does to some extent dominate the considerations of the other elements of the measurement system. This particular aspect of the cost analysis can best be evaluated by comparison of several systems that are acceptable in terms of providing the desired output. Additionally, a look at the systems already in use or available can often reduce the cost problem. If a system satisfies the measurement requirements and is not being used on another project, it is the ideal choice from a cost point of view.

In this section a numerical value of cost will not be presented, since prices seem to fluctuate by approximately 10 per cent per year. Some very general comments will be made to aid in gaining the proper perspective when confronted with a cost decision.

From Chapter 1, the emphasis on selecting the control parameters and deciding on the measurement accuracy now becomes important. The two factors that should be helpful in estimating the cost of the measurement system are the type of input signal to be measured (static or dynamic) and the number of significant figures required of the measured value. It is not always possible to obtain a system that can satisfy both of these requirements. In such cases it may be necessary to revise the initial analysis. However, by improvisation or applied measurement knowledge, it is sometimes possible to exceed, by far, the minimum requirements of the problem.

The tendency in many cases is to overdesign the measurement system. The act of overdesigning is always more expensive than that of correct design. A very approximate rule of thumb is that the increase in the significant-figure requirement by one increases the cost by a factor of five. Roughly speaking, a voltage-measuring device capable of one significant figure costs $12, two significant figures costs $60, three significant figures costs $250, and four significant figures costs $1000. If a measurement system is being purchased, strict adherence to the accuracy requirements can save a large sum of money. Only when immediate future use of the more refined systems is assured can the purchase of such a system be justified. The old saying, "You do not need an elephant gun to hunt hares," might have a valuable lesson in it. The development of a six-significant-figure measurement system for use where two significant figures are satisfactory is normally a waste of time, money, and effort. There may be one time when the use of an extravagant system can be justified. This would occur when this system was already available and not in use.

The cost of sensing elements is usually nominal. When the cost includes the price of calibration, the charge is much more. However, the major cost consideration for any system is normally the cost of the recording (read-out) devices. This is obviously not true for the measurement of temperature using a thermometer. It is true for most measurement systems that employ sensors which change an input mechanical signal to an output electrical signal. The cost for recorders for dynamic signals is much higher than for static-signal electrical recording. The selection of a sensor does partially establish the cost of the entire system, because the modifying equipment and the read-out equipment are often fixed by the selection of a sensor.

Although the cost of the measurement system may be very high, it may be very small when compared to the total experimental program. It is always important to make measurements as accurately as the system being used can measure. The greater the number of significant figures available, the greater will be the understanding of the operating conditions. When the time of the test is long and the cost of the entire project high, the measurement of an additional significant figure can be justified.

Without any exact solutions, the sensors of the previous examples will be discussed in terms of cost and in terms of the modifying and read-out systems required. When alternative modifying and recording elements are available, this will be pointed out if a large difference in system costs results. One item of cost that should not be overlooked is the cost of calibration. This cost will depend on local facilities and work load and will not be included in the following discussion.

EXAMPLE 3.5

Cost Considerations. The cup of water costs essentially nothing. The other elements might be either a light to reflect off of the surface or an electrical-resistance device placed in the cup to tell the change in the position of the water level with time.

The cantilever beam and strain gage probably cost less than $10 each. The modifying and read-out elements will depend on the accuracy and dynamics of the system. With static inputs the measurement of strain with a potentiometer-type system is sufficient for operation. When the input signal is dynamic, the modifying and recording elements could be of several types. A strain gage, Wheatstone bridge circuit, and oscilloscope could be used. This system would cost approximately $1000.

The piezoelectric system would require a charge amplifier and an oscilloscope. Cost of the entire system is approximately $2000.

The cost of beam balances can be relatively expensive, depending on the workmanship put into the knife edges. The total price often also includes a set of calibrated masses. The cost could range up to several thousand dollars, depending on the range and accuracy of the system.

Any complete cost analysis of a measurement system must be made after a definite value for the number of significant figures required has been established. The type of input signal to be measured also plays a dominant role in a correct cost estimate. Once these two parameters are known, a check in manufacturers' catalogs can reveal a cost analysis for the measurement system. Fortunately, catalog data on a given element includes complete specifications and limitations for the equipment. For additional information the local sales representative is available for consultation.

The sixth requirement that must be considered when selecting a sensor is the type of modifying and recording elements that must be used to complement the sensor. Some discussion of these elements has already been made in the sections on dynamic response and cost. The output of any sensor may be in the form of any physical variable. The sensor may change the form of the input signal to a form that is more easily modified. Sensors which also change the form of the signal input are called transducers. The cup of water senses acceleration and transduces this signal into a displacement of the surface of the water. The beam senses an applied force and transduces this force into a displacement or strain. In fact, most sensors are also transducers. The beam balance may be considered to be a complete measurement system in some cases. These systems do not transduce the input signal to another form (under null-balance operating conditions).

Two of the most common output signals of sensors take the form of a displacement or a change in an electrical property. Although it is not necessary, most modifying elements are designed to accommodate an electrical signal. Mechanical, hydraulic, thermal, and electrical systems are all equally capable of being used with static signals. The tendency toward electrical systems has resulted from the reduced friction and inertia problems associated with electron flow. Therefore, the modifying and recording elements of the measurement system will consist primarily of electrical elements. It is important to realize that other systems are available and that for static signals these systems may be much less expensive for the same accuracy than that of a corresponding electrical system.

Since the following two sections are devoted to consideration of modifying and recording elements, respectively, the subject will not be continued here. The relationships between the elements to a measurement system are realities. They must be understood if measurements are to be meaningful.

The section on sensors can be completed with a very brief outline of the techniques developed here. To select a sensor for detecting a given signal, the physical laws involving the signal to be measured should be reviewed. This leads directly to several sensor models that are capable of functioning in a sensor role. The equations also indicate any extraneous signals that might disrupt the sensor operation. The accuracy and dynamic response occupy unique positions in sensor selection. They are the final criteria used to evaluate a list of sensors for possible flaws in the system. From the list of sensors that have passed all of these previous considerations, the selection of a sensor element is made based on a comparative cost and associated modifying and recording elements. The first criterion is to prove that the sensor will perform the desired service. The second is to provide this service at minimum cost. The reader should realize the implications of these statements and should be capable of an evaluation of systems based on these criteria.

Modifying Elements Modifying elements take an input signal (which is the output of a sensor), modify this signal, and provide an output signal in a form compatible with the recording element. (See the block diagram of Figure 3.4.)

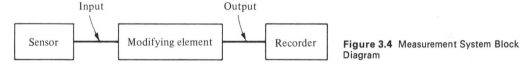

Figure 3.4 Measurement System Block Diagram

Three of the major functions of the modifying element are to change the input signal to a proper form, to filter the input signal to remove extraneous inputs, and to amplify the input signal. Every measurement system does not necessarily contain a modifying element. This function may be performed by either the sensor or the recorder. The method of separating a measurement system into functional parts is arbitrary. When it is more convenient to define a system without a modifying element, this should be done. However, a clear understanding of the function of each component of a system does aid in the analysis of the system and does provide a logical sequence for recognizing problem areas.

One of the most enlightening statements that can be made concerning modifying elements at this point is that these elements are also sensors. Therefore, all considerations made in the section on sensors are equally applicable to the present elements. The modifying elements are designed to sense the signal put out by the primary sensor element (the element that initially senses the parameter to be measured). These modifying elements are designed to be insensitive to all other input signals. The accuracy, range, and dynamic response are all designed to be compatible with the primary sensor. The output desired is that which will be most desirable for the recorder to be used.

Any techniques developed for use with sensing devices can be used in an analysis of the modifying elements. For the present discussion, the details of the systems employed in the role of modifying elements will be neglected. When the signal available from the primary sensor is not in a form acceptable to a recording device, a modifying element is required.

When specific primary sensors are considered in Chapter 6 and in successive chapters, the modifying elements available will be considered in detail. Because of the similarity between modifying elements and primary sensors, the fundamentals of operation are already known and it is not necessary to present any new basic information.

In looking ahead to the various applications of modifying elements, electrical circuits will be used most frequently. Some modifying elements that will be covered are resistance bridge circuits, ballast circuits, filtering circuits, and amplifying circuits. Most electrical circuits are easily analyzed for the six requirements on a sensor. Time and effort will be used for most of the modifying elements mentioned above, because they find use in almost every measurement system that involves an electrical form of the signal when operating. A complete analysis of these simple circuits leads directly to the problems in the control field, and a systems approach to the measurement system can be invaluable when calibration problems are to be solved.

Read-Out Devices The ultimate object of any measurement system is to provide information concerning the state, or condition, of the scientific phenomenon being investigated. This purpose can be accomplished when the final output of information is in a form that is meaningful to the observer. The output information could be the complete analysis of the system determined from a computer programmed to give output statements. This type of output is much more refined than the normal type. Usually, the output information is in the form of values of the variables being measured, along with information concerning the accuracy of the data. The scientist is then able to use this data to evaluate the phenomena under investigation.

The use of direct computer coupling to the measurement-system output device must be preceded by a calibration of the measurement system. Therefore, the fundamental problem to be considered reduces to an analysis of the components of the measurement system. The use of standards and calibration techniques will be considered in Sections 3.4 and 3.5. The present section will be restricted to an analysis of the basic types of output devices.

All read-out devices function by accepting an input signal, transducing it to another form, and presenting it in a form that is meaningful. This element of the measurement system, therefore, is functionally identical to the sensor–transducer element. Much of the operational fundamentals are again established by this statement. Of course, both the read-out device and the modifying element are specialized sensor elements, but any analysis of a sensor system can be applied directly to these elements. When the methods of calibration techniques are considered, the application to all elements of the measurement system should be realized.

Most of the discussion of read-out devices in this section is directed toward voltage-sensitive output. Since electronic equipment does play a dominant role in measurement systems, a majority of the output devices will see an input signal that is in the form of a voltage. This input signal will be transduced into some form of a signal that can be used directly. There are several types of output that are useful.

Three different types of output devices will be considered: (1) pointer–dial, (2) digital output, and (3) graphical output. To clarify this division, several examples of each type will be given.

EXAMPLE 3.6

Division of Output Devices

1. Pointer–dial examples: Clock, voltmeter pointer–dial, Bourdon-tube pressure gage, speedometer, torque wrench, platform scales, micrometer, manometer, and many others.
2. Digital output examples: Odometer (mileage meter on an automobile), yes–no light (on or off), digital voltmeter, digital resistance box, electronic counter, mechanical counter, bell timer, and others.
3. Graphical output examples: Oscilloscope, oscillograph, motion pictures, magnetic tape, recorded sound, and others.

The list is far from complete, and the divisions are not all inclusive. For instance, the purpose of the output device is to provide information that is meaningful to the scientist, and the scientist is capable of understanding outputs in the form of touch (heat, cold, roughness, etc.), outputs in the form of smell, and outputs in the form of taste, in addition to those of sight and sound. Therefore, systems may be used which do not directly employ the division of output devices previously defined. The divisions were made to encompass the major instruments used in measurement.

Pointer–dial output devices are basically static output systems. The use of the human eye to read the output immediately limits the signal to, at most, slowly changing inputs. It is possible to use these output devices for dynamic measurements, but some additional output system must be added. When the output is static or slowly changing, the basic equations governing its operation reduce to

$$A_0 I_o = B_0 I_i$$

where I_o is the output information, I_i is the input information, and A_0 and B_0 are constants. The system response is that of a zeroth-order system, because the input signal is constant. If the input signal varies, the output signal typically responds like a second-order instrument.

This type of output device can be analyzed most simply by considering the operation of the D'Arsonval meter movement of Figure 3.5. The pointer

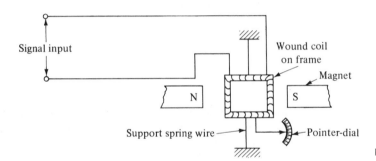

Figure 3.5 D'Arsonval Meter

is moved when an electrical field, produced by the flow of current through the coil of wires, is opposed by the magnetic field established by the magnet. The rotation is opposed by the torsional spring constant of the support wires. An analysis of the forces involved in the operation of this meter gives

$$I\ddot{\theta} + C\dot{\theta} + K\theta = F \tag{3.1}$$

where F = force of interaction between the electrical and magnetic fields
$\quad\theta$ = angle of rotation
$\quad I$ = moment of inertia of all the moving parts
$\quad C$ = resistance to motion (friction, air resistance)
$\quad K$ = torsional spring constant of the wire

The system can be designed to respond over some ranges of input frequencies, but one principal limitation is the mass of the pointer. For very-high-frequency input signals, the pointer would be deformed because of the inertial forces present in the pointer mass. The normal range of frequency is up to 100 Hz. Even at this range, some additional method of recording the data must be available, since the human eye cannot follow this transient. The use of a small mirror in place of the pointer and a light-ray source can extend the input frequency range up to 500 Hz. However, the basic device is a second-order system; the overall implication of this statement will be covered in more detail when system response to basic types of inputs is covered in Appendix III. This is the section on dynamic calibration. The frequency–gain response of this type of system is typical of the spring–mass–damper system and will be familiar to all soon.

Static voltage measurement is a very common requirement in measurement systems, and an understanding of the operation of voltmeters is important. The meter of Figure 3.5 is the basic detector in voltmeter–ammeter systems. The circuits used in association with this meter are shown in Figure 3.6. Only one D'Arsonval meter is used in each system. This meter has the

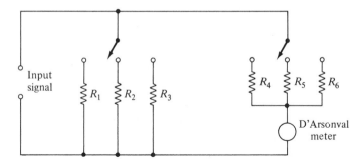

Figure 3.6 Voltmeter-Ammeter Circuit

characteristics given by Equation 3.1. It would be possible to change some of the components to allow for a large range of operational capabilities, but this is impractical. One fixed value of current through the coils will cause the maximum deflection. The two resistor-selection circuits of Figure 3.6 represent the method used to adjust the flow of current through the meter to less than the fixed maximum value. If a current flow is allowed to exceed the maximum value, the spring wire may be permanently deformed, and the coils may be melted. This is one precaution that must be observed when using this type of read-out device. The selection of the maximum value of series resistance (R_4, R_5, or R_6) and the minimum possible value of shunt resistance (R_1, R_2, or R_3) will provide for the minimum possible flow of current through the meter. This protects the meter from damage when the value of the input signal is unknown. The subsequent reduction of the series resistance and increase of the shunt resistance allow the proper selection of current flow through the meter. Knowing the value of the resistances connected and reading the current flow through the meter provide the information necessary to determine the input voltage and current.

On a real voltmeter, the selection of resistance is accomplished by using a sensitivity switch. For protection, the switch should be set so that the voltage required to produce maximum deflection (full-scale deflection) is the maximum. This is usually the 1000-V scale.

One other aspect of the voltmeter should be considered at this point. This problem is the loading imposed on the circuit being measured by the introduction of the meter into the circuit. The effective resistance seen by the input voltage determines the current flow that will be drained from the input signal. When the voltmeter does not have internal amplifiers, the loading of the original circuit can introduce a significant error. The method of rating a voltmeter is to give the ratio of resistance of the meter to voltage or the input signal. A good voltmeter having no internal amplifier has a rating of 20,000 Ω/V. An inexpensive meter might have a rating of about 1000 Ω/V. The circuit loading of vacuum-tube voltmeters is usually negligible. It is essential that circuit loading always be considered if the output information is to be reliable.

The common use of electrical measurement systems has made the measurement of voltage a principal function of read-out devices. The pointer-dial voltmeters discussed so far have all been for the measurement of direct current (dc) only. When a measurement of alternating current (ac) is to be made, some form of rectifying circuit must precede the circuit of Figure 3.6. The normal circuits vary from a simple dc capacitance bank, to a diode rectifying circuit, to a true root-mean-square circuit. It is important to measure only dc voltage with a dc voltmeter. This precaution might prevent the destruction of a D'Arsonval meter.

One other very important dc voltage-measuring device is a voltage-balancing potentiometer. This type of read-out device can measure voltages in the microvolt (10^{-6} V) range. One particularly useful measurement system requiring a potentiometer is that of a thermocouple circuit. The potentiometer is a static-input signal-measuring device only. It should not be used for other purposes unless the limitations are clearly understood. The operation and performance of the potentiometer will be considered in more detail in Chapter 6.

The digital output devices are also basically static output systems. The electronic and mechanical counters can record transient phenomena, but the visual read-out limits the signal either to slowly changing input or to sampled data. Data may be recorded for a given time period and displayed. Then data may be sampled again and a new value displayed. Both types of counters can record transient inputs. The frequency range of the electrical counter is approximately 100,000 Hz. This number depends on the frequency of oscillation of the internal oscillator. The mechanical counter is limited by friction and inertia to much lower frequency ranges (1000 Hz is high for a mechanical counter). It might be of interest to consider the operation of an electronic counter in more detail. Figure 3.7 shows the

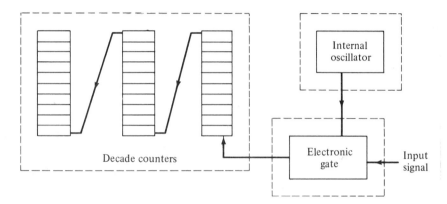

Figure 3.7 Elements of an Electronic Counter

three basic elements of the electronic counter. There are three modes of operation for the electronic counter: manual, EPUT (events per unit time or cycles per second), and PER (period). When manual gate is selected, the input signal passes through the electronic gate to the counters; each time the voltage level increases through that value selected on the triggering level circuit, the counter registers one count. When the counter is operated in the EPUT mode, a display time selector opens the gate. The internal oscillator is used to close the gate after a fixed length of time. This length of time is selected by a control knob on the counter. The input signal again is counted by the counter in the manner described before. In the PER mode, the input signal opens the gate when the voltage level is of the selected magnitude and slope and closes the gate when this same magnitude and slope again occur in the input signal. The gate passes the voltage output of the internal oscillator to the counter, where the time (or number of oscillations) is counted. The time counted represents the time between two similar points on the input signal. This is also the period of the input signal.

The selection of sampling time limits the measurement essentially to a steady-state input frequency. Any rapid fluctuations in input frequency could make the information recorded valueless.

Graphical output devices are capable of measuring dynamic input signals. All oscillographs operate using the basic D'Arsonval-meter movement, with the pointer replaced by an ink pen, a heated point, or a light beam. The dial is replaced by corresponding graph paper, thermal-sensitive paper, or light-sensitive paper, and the paper is driven by the writing instrument at various selected constant speeds. The dynamic response of the pen movement is then that of a second-order system. However, the inertia is generally larger for these systems because of the increased mass requirement on the pen. The ink and thermal pens can respond to input frequencies up to approximately 100 Hz while the light beam can respond up to 500 Hz. This is reemphasized here to establish the basic limitations of the systems. Additionally, the paper speed is not constant with respect to time and must be calibrated using a known-frequency input signal. The basic drive unit has inertia problems on starting, and there may be relative motion between the graph paper and the roller used to drive the paper. Both of these problems can be resolved by using a known input signal for calibration. If two recording pens are available, it might be possible simultaneously to record time and the input signal to provide calibration at all times. This arrangement is desirable when relatively long recording periods are to be used.

Magnetic tapes and recorded sound are essentially the same type of recorders. Very-high-frequency signals can be recorded in the form of magnetic orientation or roughness. The upper limit for this type of signal is approximately 100,000 Hz. Less expensive instruments are available when lower frequency response is acceptable. The meaning of a frequency response of 100,000 Hz is that, if an input signal of that frequency were imposed on the system, the output signal would not be distorted. If higher-frequency input signals entered, the output could not exactly follow the input.

The oscilloscope can avoid the problems of friction and inertia inherent in the other systems discussed. The electrons flowing through an evacuated tube meet little resistance, and their mass is so small that inertia can be neglected for normal input frequencies. The frequency response of the oscilloscope is high enough that practically all measurements taken with mechanical or thermal systems can be considered true. The oscilloscope is probably the best dynamic-response recorder available. It has one disadvantage in that it usually does not provide a permanent record of the output signal. Except for storage oscilloscopes, the output signal remains on the screen only as long as the persistance lasts. Additional recording equipment might be either a Polaroid camera or a high-speed motion-picture camera. Both methods are relatively expensive and often difficult to use.

Digital voltmeters, oscillographs, oscilloscopes, and magnetic tape recorders are all capable of measuring either ac or dc voltages. However, the digital voltmeter must have a rectifying circuit for operation.

All of the above systems have amplifiers built into the basic system to provide the signal amplitude necessary to drive the output device. An evaluation of amplifier circuits must be added in series to all of the measurement systems being analyzed. However, the amplifiers do not, in general, affect the response of the system.

The high-speed motion-picture camera has provided the means of recording at high speed and observing at stop action. The signal can be amplified using camera lenses and projection lenses. Advances in quality of film for these cameras have allowed very distinct object images. An internal timing light is usually provided with each camera to mark the film at a rate controlled by an external oscillator. This technique eliminates the problems associated with film acceleration through the camera. Pictures have been taken at the rate of 1,000,000 pictures per second. The dynamic response could then be as large as 500,000 cycles in some cases.

With the broad input frequency range, both the oscilloscope and the high-speed camera have been used for nearly zeroth-order systems for most input signals. Therefore, for high transient responses, either of these systems could be used to give reliable dynamic response. The magnetic tape systems can also respond to very-high-frequency input signals. If the magnetic tape is to give an output signal, the information must be transformed into a form that will be meaningful to an observer. The requirement can be met by an additional output system to convert the stored data into the form of a voltage. Then the entire system would also require another type of recording device.

Ultimately, the basic concern for a read-out device is its response to a dynamic input signal. This is also one of the primary calibration problems. Perhaps the best justification for the general response considerations is the fact that any output device capable of responding to a dynamic input signal can also usually handle a static input signal. Therefore, all output devices satisfactory for dynamic operation can generally be used in a static measurement system.

The vast majority of the measurement systems now in use employ an output device that is either a pointer–dial type or one of the voltage–type devices considered in this section. If any other type of output element is required, the fundamentals should be well enough understood to provide a reasonable acquaintance with the device.

3.4
Standards

Standards are used in the measurement system to provide a method for evaluating the accuracy and operation of the system. The particular standards used for each system will be considered when that system is discussed. The general laboratory practice for most laboratories engaged in scientific research is to maintain a relatively complete set of secondary standards. A secondary standard is a standard that has for its authority a primary standard, while a primary standard is the single authority for establishing the measure of a particular parameter.

The basic foundation for all measurements is the primary standard for such measurements. If scientists are to be able to communicate, it is mandatory that each dimensional quantity be well defined. If the definition is to be accessible to all, it must be accepted to the extent that people generally use the defined quantity to describe the situation. Therefore, the two basic requirements that must be met for a standard are that it must be well defined and it must be commonly accepted. A primary standard is the best standard available for scientific use. Primary standards are established by agreement of all those closely associated with the physical systems requiring the measurement of a variable that has no previously established standard.

The basic exchanges of material, food, and money between nations could be accomplished only by common agreement on standards of measure. The construction of roads and structures depends primarily on the establishment of standards. And the development of any scientific technology between research groups depends on the acceptance of a standard for measurement.

Therefore, every dimensional quantity that might be involved in a measurement system must have either a standard for reference or a method of derivation from such a standard. The problem of establishing standards then reduces to the needs of the scientific community and the commercial community. The requirement that a standard be well defined might be easily satisfied only by international agreement. An international body empowered with the duty to establish primary standards for measurement has been functioning with U.S. support since 1875.

The present problem faced by the international standards group is the establishment of uniform standards for all nations of the world. There are four basic dimensional systems in common use in the world. The expense involved in reducing this number to one system of measure is the primary obstacle that must be overcome before one system of measure can be established. The convenience of having a single system of measure has been recognized for some time, but the acceptance of a single system of measure has not been accomplished. Therefore, the scientist has been required to become familiar with all of the commonly accepted standards of measure. The problem of conversion from one system to another has been partially resolved by international standards conferences. However, the real solution ultimately will be the general acceptance of one system of measure. Because of the simplicity of the decimal system in the mathematical use of measurement systems, the units of meter, kilogram, and calorie are desirable. The time unit of seconds may be relatively unchanged by the present controversy, but the Kelvin temperature scale may be adopted. For the present it is still mandatory for scientists to be capable of converting from one system to another.

In the United States, the principal source of standards is vested in the National Bureau of Standards (NBS), which is administratively under the Department of Commerce. This organization establishes the standards that are to be used on all commerce. Therefore, the scientific world has relied on this bureau to establish the standards that are also used in scientific work. The basic standards for each quantity measured are maintained by the NBS, where all primary standards for the United States are available. The NBS provides a calibration certification service that can be used by any facility that might request the service. When a primary standard is not available in the laboratory, a certified secondary standard may be purchased and certified by the NBS.

The purpose of the standard in the measurement system is to provide a method of proving that a measurement is correct. Measurement is the process of comparing an unknown quantity to a known quantity. The "known quantity" in this comparative process is determined by performing the same comparative process but using a standard in the place of the "unknown quantity." The process of establishing the "known quantity" is called calibration. This is also the method employed to prove the validity of the measurement system.

3.5
Calibration

One of the most serious mistakes made by the users of measurement systems is that of accepting the reliability of the system for measurement. A thermometer hanging on the wall measures temperature, a ruler in the desk measures length, and a voltmeter sitting on the shelf measures volts. Such blind faith in these measurement systems has propagated measurement errors through negligence. One of the first responsibilities of the user of measurement equipment is to ascertain the reliability, reproducibility, and accuracy of this equipment. This can be accomplished by a complete calibration of the equipment over the range of input signals to be measured and by providing dynamic calibration if the system is to measure dynamic signals. In laboratories where constant calibration is maintained, a convenient spot check is usually sufficient. The most refined measurement equipment available should not be used unless it has been calibrated. The individual who is knowledgeable in the field of measurement will always be aware of the limitations of any data measured using uncalibrated and unfamiliar equipment.

The calibration of a measurement system is normally accomplished by using a source signal which is either a primary or a secondary standard. If it is possible to obtain a variety of input magnitudes, a graphical plot of input versus output can be used for the calibration curve in the case of static systems. An ideal measurement system for static signals would obey the equation

$$A_0 I_o = B_0 I_i$$

where A_0 and B_0 are constants, I_o is output information, and I_i is input information. Here there is a linear relationship between input and output information in the measurement system. The major convenience of knowing that there is a linear relationship is that static calibration can be established with fewer points than might be required for more complex relationships. However, similog and log–log paper or variable grouping can also be used in many cases to obtain straight-line plots. These methods will be presented in more detail in Chapter 5.

When all of the signals to be measured are changing very slowly (essentially constant for a particular operating condition), static calibration is the only basic verification required of the measurement system. However, if the input signal to be measured will change in value during the time of measurement, the measurement system must be calibrated using a dynamic input signal of some type. Dynamic calibration presents two major problems: first, the output information cannot usually be plotted against the input information; second, standards for use in calibration are seldom dynamic inputs. The first problem can be resolved by obtaining the differential equation relating the output information to the input information. This is the most complete calibration possible under the circumstances. The second problem must be approached by the fabrication of input signals which are composed of standard values in some way. Both of these topics will occupy a fairly lengthy discussion. To avoid the confusion involved in a lengthy discussion of calibration theories and techniques, a discussion of the details of calibration appears in Appendix III. The presentation of calibration concepts concludes the present discussion of measurement systems; the subject will be considered in more detail in Chapter 6 and the following chapters. The last item in experimental analysis, data analysis and project evaluation, will be introduced in Chapter 5.

3.6
Conclusions   A detailed study of the measurement system has been delayed until Chapter 6. The present chapter presented the basic elements, functions, and fundamentals of the design of a measurement system. The principal material to be remembered can be summarized:

1. The measurement system is to determine a quantitative value of a parameter to be measured.
2. The act of measurement changes the system being measured.
3. The elements used to accomplish the comparative measurement are sensing elements, modifying elements, and read-out devices.
4. Every measurement system must be calibrated before the system can be used to measure variables.
5. All calibrations must involve the correct primary or secondary standard to be valid.

Much more of the details of the calibration and operation of specific measurement systems will be covered in later chapters. The present chapter merely established an understanding of the fundamentals of measurement. These fundamentals will be employed when each new measurement system is designed, and the information obtained in this chapter will find constant application throughout the remainder of this book.

PROBLEMS

3-1. A mercury-in-glass thermometer is used to measure temperature. Determine the physical law which governs this instrument. Discuss any extraneous inputs that could cause an error in the measured temperature. Give an engineering judgment of the response of a thermometer.

3-2. Define three measurement systems capable of sensing pressure. Give the physical laws governing their operation.

3-3. Design a measurement in which a thermometer can be used to read pressure.

3-4. A thermometer reading 68°F and weighing 2 oz is placed in a thermos bottle containing $\frac{1}{10}$ oz of ice and 1 oz of water. What will be the steady-state temperature measured (assume the thermometer has the same properties as water for this problem only)? How does this relate to the first axiom of measurement?

3-5. Consider one property the human body can sense. What extraneous inputs complicate the measurement? What is the response time of the human measuring system?

3-6. A stainless-steel ruler (1 ft long) can sense temperature. (Tell how.) Would a pressure change of 100,000 psi introduce any error in the measurement systems?

3-7. Give two sensors capable of detecting humidity of the air. How do these sensors obey the first axiom of measurement?

3-8. What is the practical operating range of a thermometer? (Why?) What limits the accuracy of the thermometer?

3-9. Break the components of a radio into three parts: tuning circuits, amplifying circuits, and speaker. Tell exactly how the radio compares to a measurement system.

3-10. What is the response of a radio (or stereo) to the input signal? How does the radio obey the first axiom of measurement?

3-11. What are the modifying and recording elements on the thermometer?

3-12. Consider a thermocouple, dc amplifier, and oscilloscope for measuring temperature. Discuss the response of each element of the system.

3-13. What determines the dynamic response of a digital counter. What information about the counter provides an absolute bound on the most accurate reading possible with the device?

3-14. How could you find the dynamic response of an oscilloscope? An ink-writing oscillograph?

3-15. What two characteristics limit the dynamic response of a simple D'Arsonval meter with pointer and dial?

3-16. What are the limitations on the dynamic response of high-speed cameras?

3-17. How is measurement related to calibration? How is calibration related to standards?

THE ENGINEERING PROBLEM

4

4.1 Introduction

Now it is possible to establish a basic approach to the solution of engineering problems. The information contained in the first three chapters was not generally detailed enough to relate it to a specific problem. Therefore, the present chapter will attempt to provide sufficient examples for the application of this information to a problem solution. However, many of the problems in data analysis and actual measurement-system selection can not be solved at this point. Some reference to material in Chapters 5–9 will be required when the problem-solving format suggested in this chapter is applied to a different problem.

The purpose of any experimental analysis of a physical problem is either to verify a mathematical model for the phenomena or to establish such a model. The ideal model would be in the form of a simple functional relationship between the variables. However, a table of values or an engineering plot of data are both usable models for a particular system. To accomplish the purpose proposed above, a format should be available for use on any engineering problem. This format would include a step-by-step approach to the problem, including the questions that must be answered at each point in the problem solution.

One possible format has generally been presented in the preceding chapters of this book. However, we should be able to attach more meaning to the directing statements now that more experience has been gained. Therefore, an attempt will be made to establish this format in detailed form and to use it in several example solutions to general engineering problems. It will not be possible to answer particular questions that might arise in using the format, but the examples have been selected to present a variety of engineering problem areas. The method of approaching a problem is a critical part in the types of solutions obtained. We would like to have a format that allows a variety of choices in the pursuit of a solution. The problem format to be recommended should enhance either the speed of solution or the quality of solution (or both). In all cases the format is not to be an absolute boundary obstructing the creative ability of the engineer. Rather it is to be a guideline to maintain direction and progress in the experimental program.

4.2
Format for Engineering Problems

A problem must be properly recognized and classified before it can be well defined. This is not always easy to accomplish, but it is the essential first element that must be considered in any application of experimental analysis. We will list a format that could be followed in a problem solution. Then, general comments will be made on the applicability of the format to a problem. The types of questions that should be asked at any point in the format will be discussed. The flow chart that will result from the answers to these questions will plot the course of action to be taken.

GENERAL PROBLEM-SOLVING FORMAT

Part 1. Obtain a complete definition of the problem.
    (a) List the principal goal sought.
    (b) List all secondary goals.

Part 2. Perform a complete literature review.
    (a) Establish the models known to apply to your problem for given ranges.
    (b) Determine if there are any areas where incomplete information exists.
    (c) Define the problems that must be answered to attain the goals of Part 1.

Question 1: Is enough technical knowledge available to attain the basic goals?

Answer 1:   Yes
    (a) Perform necessary design calculations.
    (b) Build the system or a model and test the system or model for performance.
    Go to Part 6 for test.

Answer 1:   No
    (a) Isolate the problem areas.
    (b) Approach each area in the following steps.

Part 3. Define the boundaries of each problem area.

    (a) Draw a control volume around each problem area and define the variables that will be inputs and outputs of this volume.

    (b) If possible, write the general equations relating the input variables to the output variables. If no mathematical model is obvious, try an analog model or use dimensional analysis to obtain the important dimensionless groups.

    (c) Determine all parameters that must be measured to evaluate input–output information.

    (d) Approximate the accuracy with which each parameter must be measured.

    (e) Select the measurement systems that can satisfy the above requirements.

Part 4. Design an experimental program to obtain the relationship between the input and output information of Part 3.

    (a) Select the input parameters (control variables) and the output parameters.

    (b) If a mathematical model is proposed, test it.

    (c) If an analog model is proposed, test it.

    (d) If dimensional analysis is used, establish the relationship between dimensionless groups.

Part 5. Perform an analysis of the data obtained in Part 4.

    (a) Obtain a least-squares data fit for all models.

    (b) Perform an error analysis of the measurement systems.

Question 2: Does the model satisfying Parts 3, 4, and 5 provide the information required to attain the basic goals of the design?

Answer 2:   Yes

    Go to Answer 1: Yes.

Answer 2:   No

    (a) Check to see that the accuracy of measurement is satisfactory.

    (b) Analyze the problem for other important parameters.

    (c) See if any unknown extraneous inputs are present.

    (d) Test the feasibility of the proposed model.

    Go to Answer 1: No.

Part 6. Test the final product (or model) under normal operating conditions.

    (a) Draw a control volume around the final design product (or model).

    (b) List all input parameters and output parameters.

    (c) Establish the accuracy with which each parameter must be measured.

    (d) Select measurement systems for each parameter which will satisfy the requirements of (b) and (c).

    (e) Apply the required input information and test the product response.

Question 3: Does the designed product (or model) perform in such a way as to satisfy Part 1 ?

Answer 3:   Yes

Write the final report with final design and limits of accuracy and boundary conditions.

Answer 3:   No

    (a) If possible, isolate the portion of the design that does not respond properly.
Go to Part 3.

    (b) If the problem area cannot be isolated, check the measurement systems for extraneous inputs and accuracy.

    (c) Check the total problem for inputs that were not specified in the original design.

    (d) Perform an error analysis on the data collected.

    (e) If none of the above lead to the source of the problem, go to Part 1.

    (f) If the problem area is determined,
go to Part 3.

DISCUSSION OF THE FORMAT

Parts 1 and 2 are presented in more detail in Chapter 2, which should be reviewed now. It is always important to list the overall goal and all secondary steps that must be taken to attain that goal. This gives direction to the experimental program and provides a logical sequence for approaching the problem. A review of the present state of the technical knowledge is important in avoiding an experimental approach to a problem that has already been solved. We tend to attack a problem with much enthusiasm; this can lead to an incomplete literature review, because we wish to "do something" about the problem. We may, therefore, spend several weeks duplicating work that has already been done. A few days spent performing a complete literature search can save several weeks' work.

Question 1 should now be answered. If the technical knowledge is adequate for attaining the basic goals, the knowledge should be applied to the problem and a design product should result. If the final installation is a large one, it might be economical first to build a model for testing before the final product is built. An introduction to some modeling techniques is given in Chapter 2. If the final installation is relatively inexpensive, it may save time and money to build the final product and test its operation. In either case the final design must eventually be tested under operating conditions.

If the answer to Question 1 is no, the areas where technical knowledge is lacking should be listed. These areas must all be explored to determine the importance they will have on the final design. Frequently, the problem can be overwhelmed (by designing a product with much more capabilities than are required) at less expense than it would take to resolve the unknown areas.

In Parts 3, 4, and 5, a basic approach to experimental problems is given. This might be called a systems approach, a black-box approach, or a thermodynamic approach with equal applicability. In any case a problem is approached by defining a boundary that contains the system under analysis. Any inputs which cross this boundary must be considered in the analysis. The outputs are the response of the system to these inputs. In thermodynamic considerations, the inputs and outputs are energy and mass. In some cases we are interested in only one particular form of input and output such as velocity, displacement, voltage, and so on. A complete thermodynamic analysis may not be necessary if the order of magnitude of one term dominates the problem.

It is usually desirable to predict a mathematical model for a problem area if one has enough technical skills to do so. However, an analog model or dimensional analysis can be used to obtain the same type of information. Appendix II presents the methods of dimensional analysis. All important parameters should be measured with an accuracy consistent with the acceptable error limits on the design. The selection of the proper measurement system is covered in Chapters 6–9. This is not always an easy task, but it is an essential one. All measurement systems should be calibrated in the range of their use and for the type of input signal to be measured. Their operation must be understood if the output information is to be meaningful.

The design of an experimental program to obtain the relationship between input and output information is much more involved than might first be expected from the statement. Selecting the input parameters and controlling these parameters may involve the design of several other systems.

The range of the test should extend beyond the range of the final product if this is possible.

The relationship between input and output information should be fitted with some mathematical function if this is possible. A least-squares technique will be presented in Chapter 5. This type of data fit is desirable for establishing the limitations for the results. It can also help to discover if extraneous inputs are present. If all of the parameters which influence the problem have not been listed, they could appear in the form of an extraneous input or a biasing error.

An error analysis of the measurement system can be applied to the above-determined mathematical function to establish the portion of the variation that could be the result of measurement error. If the possible measurement error exceeds the observed experimental error, it might be necessary to select more accurate measurement systems.

If the answer to Question 2 is yes, we are ready to design the final product. If the answer is no, we must refine the experimental program. Again we look at the accuracy of the measurement system and for possible extraneous inputs in its operation. We may have to include some of the parameters that were previously thought to be insignificant. It may be necessary to list each component part of the problem area and consider each one individually in an experimental program. Error sources are usually easier to isolate at this point in a program than in the final design. It may also be necessary to start the entire process with a different point of view, or a different method of accomplishing the purpose may be sought. If the problem cannot be solved in the time allowed, we may have to attempt to compensate for this lack of technical knowledge by overdesigning the problem. This may not always be possible, and it is generally a more costly method of solution. Therefore, it is less desirable to rely on this approach.

The test of the final product is accomplished in essentially the same way as the experimental problem solution. The only difference is that the control volume now includes all components of the design. The answer to Question 3 is the final criterion on which the total design is evaluated. If the product satisfies the requirements listed in Part 1, a final report will be prepared. If it does not perform in an acceptable manner, we need to isolate that part of the design that causes the problem. The techniques used to find the problem area have already been discussed. If the problem area cannot be found, overdesign may be attempted at this point.

With the time and cost limitations, all problems are not solvable. Advances in technology require both. If problems do not violate known physical laws, it is expected that they have a solution. However, our present technology may make a solution impractical at this time. It is fortunate if we have the foresight to recognize this type of situation before we attempt to work on the problem. Great new discoveries come from those who will not accept failure. There is certainly a thin line separating the extremely difficult problems and the "impossible" problems (for the present).

4.3
Engineering
Problem I

The first problem to be considered using the format proposed here is the design of an air-conditioning system for a particular office building. Individual control is to be provided for each room. The problems in this chapter were selected to demonstrate different paths in the flow chart included in the problem-solving format. This first problem demonstrates one possible sequence of analysis.

The principal goal is to maintain each room in the office building at a temperature and humidity that is comfortable for the occupants. Each room is to have a control for selecting the desired temperature.

The secondary goals that must be attained to accomplish the principal objective should also be listed. This list might include the following:

1. Establish the maximum heat load for each room.
2. Determine the maximum humidity load on each room.
3. Use these to determine the refrigeration capacity required for the entire building.
4. Determine the maximum air volume flow rate for each room, based on the heat and humidity load.
5. Design the basic fan and ducting requirements.
6. Select the operating temperature and dimensions of the heat exchangers.

In the course of a literature review, references 1, 2, 3, and 5 are selected. The models have been established in these references, and the goals can be attained by applying these principles to the problem. Therefore, the answer to Question 1 is yes.

It is possible to make calculations of the maximum heat load for each room by predicting the most adverse conditions that might be encountered. These calculations will be based on the highest outside temperature and humidity that could be expected in this region, the building material and dimensions, and the most demanding operations that can be expected in each space. These conditions will establish the refrigeration required for the entire building. (The heating problem will not be discussed in this presentation.)

The volume flow rate required for each room is established from the physiological principles of references 1 and 2. The calculations depend on the building and will not be made for the present analysis. The overall requirements and calculations can be made based on all the information concerning building design, location, and functions in the space. The method of solution is well documented and the details are involved. Therefore, we will assume that they have been made and that all design requirements are met.

The final test of any design is whether or not it accomplishes the desired goals. This is accomplished by building and installing the facilities and measuring the room conditions in each space. Usually, the ducting system will be designed to provide a flow rate that exceeds the demands made on it. A test of the ducting system is given in reference 4. If the rooms cannot be maintained at the desired conditions using the design, either more airflow capacity must be provided or lower air temperature must be possible.

In this case the methods of design are well established. The only variation might come from errors in the basic calculation, errors in the construction, or errors in the functions performed in the room. It may be possible to avoid most of these problems by incorporating a safety factor in both the air-moving capabilities and the refrigeration capacity (this is overdesigning the system). In most engineering designs a safety factor is incorporated in the design which allows for 10 per cent more capabilities than were calculated. This is a concession to the fact that all variables may not have been established absolutely. If the variables are not well defined and certain, higher safety factors may be required.

The heat load and the humidity load in each space are strongly dependent on the functions that may occur in a space. If any possible variation from design conditions may exist, the system should be designed to handle these more demanding conditions.

The testing of the final product for this first example is actually relatively simple. The occupants of each space could be consulted concerning their comfort. A more complete test could be accomplished by measuring the temperature and humidity under operating conditions. Usually, cooling systems are manufactured in terms of tons of refrigeration required, and a system is selected which is larger than the calculated load and the safety factor. This actually introduces an additional safety factor in the design. The final test will prove satisfactory if errors have been eliminated. The problem is, therefore, completed.

4.4
Engineering
Problem II

The flow diagram for Problem I included Parts 1, 2, and a yes answer to Question 1. Then, Part 6 was covered to complete the analysis. Now let us look at a problem where the answer to Question 1 might be no. Reference 6 is an extremely good example of the techniques required to design an experimental program. Most of the principles presented in this textbook have been applied to the problem of measuring tread wear of tires. (The solution of reference 6 was accomplished before this textbook was written.) Let us assume for the moment that reference 6 was not found in the literature review and that we are to use the format proposed in this chapter to obtain a realistic method for comparing various tires for wear. The proposed method will be a very brief summary of the material contained in reference 6.

For Part 1 we list the principal goal of the experimental program to be: "Develop an experimental program for comparing the tread wear for different tire brands." Secondary goals would include (1) a standard method for measuring tire wear, (2) a statistical model for the experimental program, (3) an investigation of extraneous inputs, and (4) a method for predicting tread life.

For Part 2 we performed a relatively complete literature review (but we did not find reference 6) and obtained a relatively important knowledge of many of the parameters which influence tread wear. A knowledge of measuring techniques is also available. However, it is necessary to consider the overwhelming number of variable inputs in much more detail to establish a meaningful experimental program. Therefore, our answer to Question 1 is no.

Since we intend to look at all of the problem areas, we will use the literature review for basic knowledge. However, we will attempt to accomplish all of the secondary goals independently of previous work when we can find better solutions. We will first look at the problem of how tread wear can be measured.

In Part 3 the control volume is the tire and the measured values are associated with the tire dimensions and mass. Two very obvious methods of measuring wear were discovered in the literature review. One involved measuring the change in the mass of the tire, while the second method involved measuring the change in tread depth at a number of selected points around the tire. The first method (using mass measurements) will automatically give an integrated wear measurement for the total tread area, provided the tires do not run into curbs and wear on the sides as well. The measurement of tread depth could be more important, since a tire is assumed to be worn out when the tread depth is zero. However, many tire-tread design factors will be extraneous inputs to this type of measurement. This latter method is a commonly accepted method of measuring tread wear, and it is selected for use in the present experiment. We can also measure the mass loss at the same time with very little increased cost and may choose to do this as well. One method for presenting tread wear data is in terms of mils of wear per 1000 miles. This could be found from the measured data by dividing the change in tread depth by the number of thousands of miles traveled during this change:

$$\text{wear} = W = \frac{\Delta \text{ tread depth}}{\text{miles traveled}} (1000)$$

Parts 4, 5, and 6 need not be completed for this secondary goal at the present time. If the method of measuring wear should also include other parameters such as contact area, this may be determined when the experiment is run.

It is difficult to consider a statistical model for an experimental program without also looking at the control variables. Therefore, the extraneous inputs and the control we attempt to maintain over them will be considered along with the statistical model. In Part 3 we attempt to list the input variables which influence tread wear and are faced with an extremely complex problem. A partial list of factors affecting tread wear includes road conditions, air pressure in tires, automobile characteristics, wheel balance and alignment, load on each tire, driver characteristics, road speed, environmental temperature, humidity, wind direction and velocity, road geometry, wear in automobile parts, tire mounting on the wheel, wheel mounting on the automobile, and so on. We should recognize at the outset that we do not know the mathematical relationship between these many control parameters. We will not be able to obtain absolute control over the problem. When we think of the extraneous inputs which affect the driver characteristics alone, we are unable to say how the driver will feel at a particular time.

This discussion might lead us to think that no acceptable control method can be found for accomplishing the principal goal (and this conclusion may be appropriate). However, we can devise a method for control that will eliminate many of the biasing errors that result from the above input variables. One such method is proposed in reference 6. This method was chosen to provide a statistical control of the input variables. The authors proposed that a convoy of automobiles be used to eliminate many of the input variables. They also proposed a method of rotating tires from automobile to automobile and to each position on the automobile. The drivers were not changed and equal mileage was driven with the tires at each location. There may still be some question about how the constant-mileage route on a dry, hot day compares to that for a wet, cool day. The tire at an adverse position on a dry day might wear more than a different tire at that position on a wet day. This type of error is partially resolved by using a somewhat random process for selecting the tire-rotation position. The cost of this experimental program, using 4 cars and 16 tires for 500 mi at each position, is approximately $10,000, according to reference 6.

Reference 6 also points out that the use of a geometric average of the individual wear measurements helps to eliminate some of the extraneous inputs. This method is then selected for the statistical model for the tests. Sources for this analysis are given in the bibliography of reference 6. With the mathematical model proposed, we have essentially performed Part 4 of the format. In Part 5 we must actually perform the designed experimental program before we can obtain the data. The least-squares fit is not appropriate, since the test is basically a statistical control problem.

The answer to Question 2 is a conditional yes. We may never obtain an absolute yes for this question, since the cost of such a program may not be justified by the results. We might be able to provide much more rigid controls to eliminate all extraneous inputs, but the tests would be less applicable to actual road use. With a conditional yes to Question 2, we return to Parts 1 and 2. Then the answer to Question 1 is yes. The experimental model is constructed, and the final test is presented in Part 6.

In reference 6 the experimental model was used to test a variety of tires and the analysis of the data indicated that the model was practical. This does not necessarily mean that the proposed model is the best possible model nor that it is the best model that has been proposed. However, it is a workable model, and it does provide output information that can be used realistically to compare two different brands of tires. The authors of reference 6 used a very logical method to obtain a workable experimental testing model. Their results could be used to predict tire life with reasonable accuracy.

Many of the very important details of the analysis and of the planning of the experiment were not discussed here. It is recommended that the reader obtain a copy of this reference and observe how the basic principles presented in this textbook can be applied to a real engineering problem. Again it is pointed out that reference 6 is an effort completely independent from this textbook. It was discussed here only becaused the authors (Spinner and Barton) chose a solution technique that is very similar to the one proposed here.

**4.5
Engineering
Problem III**

The first problem was a typical engineering design for a particular application. The second problem was the design of an experimental program capable of evaluating a selected parameter. Now we look at a problem that requires the design of a measurement system to determine a transient parameter. Specifically, this problem is to provide for the measurement of the surface temperature inside a shock tube.

The principal goal is to select a measurement system to determine the transient surface temperature for the passage of a shock wave. Secondary goals might include the following:

1. Seek a temperature sensor to measure surface temperature.
2. Select a sensor with dynamic response 10 times that of the surface, if possible.
3. Select the intermediate circuits and recording devices.
4. Calibrate the measurement system.

In the literature review of Part 2, we immediately reject several types of sensors for various reasons. The thermometer cannot measure surface temperature, it has relatively long response times, and the normal read-out device is inadequate for dynamic signals. All sensors with large mass will essentially be neglected, because they are unable to measure surface temperature. The thermistor and the thermocouple both offer some possibilities of approximating surface temperature. However, additional search shows that both of these sensors can be made using relatively new construction methods.

It is now possible to construct thermocouples to measure surface temperature by mounting the thermocouple wires through a hole in the material and extending them through the surface. Those parts extending above the surface are ground off, and the leads are isolated from each other electrically at this point. The leads can then be connected by one of several processes:

1. An electrically conducting paint can be applied to the surface. Here the introduction of an intermediate metal can be eliminated by painting with a material of one of the leads.

2. The surface can be placed in an evacuated space where the pressure has been reduced to less than 10^{-6} torr. Then, either vapor-plating or sputtering methods can be used to connect the leads with a material of one of the leads. In all of the above cases, the thermocouple junction is located across the diameter of the lead dissimilar to the material used in the connecting process.
3. Soldering or brazing can be used to connect the leads. This method involves relatively high temperatures and can lead to a breakdown of the electrical insulation.
4. A *totally* printed-circuit type of method can be applied on the surface using dissimilar metals.

It is also possible to construct thermistor-type sensors in much the same way as in the printed-circuit technique. Both types of sensors can be made very small at the point where the temperature is sensed. Both would seem to be very good possibilities for solving the problem suggested here. Certainly, an analysis must be made of the heat-transfer problem associated with the point where the temperature is to be measured. This is true for all temperature sensors. This information may not be available for the particular case under consideration. Therefore, the preliminary answer to Question 1 is no.

In Part 3 the boundary of the problem is selected to be the total mass of the junction where the temperature is measured. The inputs to this boundary are heat transfer from the fluid in the shock tube to the junction, the heat transfer by conduction from the junction down the wire leads and through the material of the surface, and the net heat transfer by radiation. If the sensor is flush with the surface fluid, stagnation does not occur at the junction, but fluid friction with the surface may be significant. In a shock tube the fluid friction is generally very small compared to other terms and is neglected here.

One mathematical model for this problem is to write the energy equation for the junction. This would give

$$\text{net convection}_{in} + \text{net conduction}_{in} = \text{energy stored in the junction}$$

The net convection to the junction is approximately equal to the product of a convective film coefficient times the junction area times the temperature difference between the fluid temperature and the junction temperature, $hA_j(T_f - T_j)$. The conduction to the junction can result from several flow paths. There is heat conducted down the lead at the junction, heat down the electrical insulating material between the thermocouple lead and the surface material, and heat along the surface to the other thermocouple lead and to the material of the surface. However, if the two thermocouple leads have approximately the same thermal conductivity, k, as the surface material, the heat flow can be approximated by assuming that the material is homogeneous. Also, heat transfer is generally slow compared to the speed of a shock wave, and the transient condition will exist for only a short time. For this case the conduction is approximated by a one-dimensional heat flow down the thermocouple lead. (Conduction $= kA_w \, \partial T/\partial x$, where k is the thermal conductivity of the lead, $A_w = A_j$, and $\partial T/\partial x$ is the temperature gradient down the lead.) In this expression a negative temperature gradient would remove heat from the junction.

The energy stored in the junction is the product of the junction mass, the specific heat of the junction material, and the time rate of change of the junction temperature, stored energy $= mc \, \partial T/\partial t$. The mathematical model is a first-order partial differential equation. This model can be checked if the initial conditions and fluid temperature are known.

Part 4 is used to test the model (calibrate the measurement system). A step-input change in temperature can be approximated by subjecting the surface to a high-velocity jet of water with a constant temperature (ice water or boiling water). A series of ramp-type inputs could be obtained by chopping a high-energy radiant source at fixed and known frequencies.

For one of the experimental programs selected in Part 4, an analysis of the data is made according to Part 5. An error analysis between the experimental observations and the proposed mathematical model will be completed. It is essential that enough calibration data be collected to provide for a complete statistical analysis. Since this calibration procedure will be a determining factor in the usefulness of the measurement system, the errors should be understood and explained before the system can be used with confidence in the actual test.

It should be pointed out that the approximation which neglects the conduction away from the thermocouple junction reduces the above mathematical model to the familar exponential decay problem (the response of a first-order system)

$$hA_j(T_f - T_j) = mc \frac{dT_j}{dt}$$

with a solution

$$(T_j - T_f) = (T_{j_0} - T_f)e^{-hA_j t/mc}$$

where T_{j_0} is the initial junction temperature.

If the answer to Question 2 is yes, the design of the measurement system is probably completed. It is still possible that the use of the measurement system with different fluids and under different input conditions may prove inadequate. However, the experiment proposed in Part 4 should have been designed with these factors in mind.

If the answer to Question 2 is no, the source of error may be in any of the energy terms of the mathematical model. The convective film coefficient can vary, depending on fluid turbulence, surface roughness, viscosity, and many other factors. The thermal conductivity may depend very strongly on the electrical insulating material and the variation in thermal conductivities of the different materials involved. The mass of the junction may be very difficult to determine, and the specific heat may depend on the method of constructing the thermocouple junction. The problem may, therefore, be two or three dimensional in the space variables. In fact, this particular thermocouple geometry may not lend itself entirely to a mathematical model. The measurement system could still be calibrated for use in the measurement of transient surface temperature across a shock wave if some standard shock wave sources could be made available for calibration purposes.

4.6 Conclusions

The problem-solving format proposed in this chapter can be used effectively in the analysis of experimental problems. Each problem has its own peculiarities, and the emphasis placed on each part of the format will be governed by the particular problem. Parts 1 and 2 should be performed for every problem. Parts 3, 4, and 5 incorporate the fundamentals of the experimental solution of problems. The selection of some type of model and the design of the experiment require considerably more details and work than might be expected from the guiding steps. The knowledge to accomplish these parts must result from an adequate technical background in the field of study.

The design of measurement systems is a feature of this book. Engineering Problem III points out that a mathematical model is not essential for the system to be capable of measuring. However, it is extremely useful when the questions of extraneous inputs must be answered. The final design of an experiment is strongly influenced by the capabilities of the measurement systems.

In Part 5 it is always important to remember that an analysis of the experimental data is essentially always a statistical analysis. The contents of Chapter 5 and the associated references should be the basis of any final data analysis. Enough measurements must be planned to provide the information for a statistical analysis of any experimental program.

Part 6 is the final test of all previous work. If the program was complete and correct, the final test should be successful. Part 6 corresponds to this chapter in this book. If the technical and philosophical information contained in the whole book is adequate, this chapter should contain the material to focus this knowledge on an engineering problem. The engineer should be able to divide a problem into its component parts, solve each part, and recombine them for one solution to the problem.

References

1. *Heating Ventilating Air Conditioning Guide*, vol. 36, American Society of Heating and Air-Conditioning Engineers, New York, 1959.

2. *ASHRAE Guide and Data Book*, *Fundamentals and Equipment*, ASHRAE, Inc., New York, 1963.

3. KREITH, F., *Principles of Heat Transfer*, 2nd ed., International Textbook Company, Scranton, Pa., 1965.

4. *Standards, Definitions, Terms and Test Codes for Centrifugal, Axial and Propeller Fans*, Bulletin no. 110, 2nd ed., Air Moving and Conditioning Association, Inc., Detroit, Mich., 1952.

5. JORDAN, R. C., and PRIESTER, G. B., *Refrigeration and Air Conditioning*, 2nd ed., Prentice-Hall, Inc., Englewood Cliffs, N.J., 1956.

6. SPINNER, S., and BARTON, F. W., *Some Problems in Measuring Tread Wear of Tires*, NBS Technical Note 486, U.S. Department of Commerce, Washington, D.C., 1969.

5
DATA ANALYSIS

5.1
Introduction   Data analysis is the basic vehicle by which an experimental program can be evaluated. This knowledge is necessary if the complete analysis of an experimental program is to result in a correct decision-making process. The decision might be to investigate certain aspects of the problem in more detail, it might be to terminate the project, it could be a decision to develop a product, or it could be a conclusion that the program has revealed possible solutions to many related problems. In any case the decision is based on the final analysis of the program. To prepare for the final presentation of information, the engineer must be able to document the results of a program with information that is meaningful. The most commonly accepted form of reporting information is to employ a statistical analysis of the program.

If the experimental program provided results that were always predictable, the analysis would be an extremely simple job. This condition is one endpoint in the boundary of possible experimental results. However, the most complex conditions in analyzing the program exist when the experimental design, every measurement taken, and all implications of this data must be evaluated. Under these adverse conditions, the techniques of data analysis thrive.

To become competent in the analysis of all data collected, some mathematical tools must be available. It will also be important to develop a technical language, so that communication will become concise and precise. Once the tools and the language have been developed, a variety of specific techniques will be applied to measurement data, and forms of error analysis will be considered. Calibration of measurement systems will be used for the examples initially, because this is a beautiful illustration of the principles of data analysis.

Understanding the fundamentals of data analysis will lead directly to an appreciation of the wide applications of these concepts in the analysis of measurement systems and experimental programs. While most of the fundamentals are not developed here, they are introduced and their applications to real situations are stressed.

A major goal of this chapter is to acquaint the reader with the methods, logic, and tools of data analysis to such an extent that the application to different situations is possible. A secondary goal is to emphasize the relationship that exists between the final analysis of an experimental program, the design of the program, and the design of the measurement system. Another goal is to stress the fact that a decision must be made at this point in the experimental program. This may start a new cycle or end the program.

The best-designed experiment and measurement system can become worthless if the final analysis is not convincing. The best way to present a final analysis is by using the tools and language of this chapter. These are the commonly accepted "proofs," and strict adherance to these methods usually saves time for the writer and reader. This is not intended to exclude other methods, but statistical methods have become so prominent that the scientist must understand the fundamentals in order to read other presentations.

5.2
The Language Problem

It is always difficult to memorize definitions because of the lack of motivation. It is also very difficult to sit in a discussion where the only language is Choctaw (unless you speak Choctaw). In some cases the definitions presented in this chapter are not commonly accepted, but they are used consistently in this book. The concepts, however, are commonly accepted, and it will not be difficult to use the discussions here in understanding other presentations. Therefore, motivation for memorizing the definitions presented here include both the need to understand this material and the need to be able to speak with others in this area. The most difficult problem in any first encounter with a new body of knowledge is the lack of communication. The only solution to this problem is the development of a good vocabulary. Therefore, definitions in both measurement and statistics will be introduced for application throughout this book.

The preliminary analysis of an experimental program begins with concern for the measurements taken. If two capable people used the same 1-ft ruler to measure independently the length of a wall, it would be almost spectacular if they both obtained the same number. This evidence immediately leads one to the realization that one of the measurements is incorrect. With only slightly more insight into the problem, it becomes apparent that both numbers are probably incorrect. This is no reflection on the two individuals making the measurement, because if 10 individuals made the same measurement, 10 different numbers would most probably result. This simple experiment points out a basic truth of measurement: all measurements are susceptible to error. Three basic terms are used to classify this error: accuracy, precision, and bias.

Accuracy is associated with the difference between a measurement of a parameter and the true value of the parameter. When only one measurement is made, the accuracy is simply the value measured minus the true value. Here the value measured is the output of the measurement system after any necessary corrections (probably resulting from system calibration) have been applied. For the example of 10 independent measurements of the length of a wall, the concept of accuracy involves establishing bounds for the error. This is accomplished by calculating the accuracy of each of the 10 measurements. Then the accuracy of the measurement is given in terms of the maximum of the absolute value of the accuracies determined from all readings. Typical forms of presenting accuracy result from the following word equations:

accuracy (upper limit) = maximum measured value − true value
accuracy (lower limit) = minimum measured value − true value
accuracy = maximum of [accuracy (upper limit)] or
[−accuracy (lower limit)]

Based on these definitions, the accuracy could be used (for the limited data) to predict the true value of a parameter based on a measurement:

true value = measured value ± accuracy

Suppose the 10 measurements of the wall were made and recorded. For illustration of the terms being defined in this section, the readings are taken to be those listed in Table 5.1 where the true value is assumed. Based on the definition of the previous paragraph, the accuracy is 1 ft 3 in. In equation form,

true length = measured length ± 1 ft 3 in.

This is, of course, true for the data presented; however, it is somewhat inadequate in that the description does not reflect all the information provided by these measurements.

To provide a more complete analysis of the data collected, the term pre-

TABLE 5.1   Typical Results of Measuring Wall Length with a 1-ft Ruler

Individual	Length Measured	True Length	Accuracy
1	14 ft $3\frac{11}{16}$ in.	14 ft 6 in. (4.4196 m)	$-2\frac{5}{16}$ in.
2	14 ft 4 in.	14 ft 6 in.	-2 in.
3	13 ft 3 in.	14 ft 6 in.	-1 ft 3 in.
4	14 ft $3\frac{1}{4}$ in.	14 ft 6 in.	$-2\frac{3}{4}$ in.
5	14 ft $3\frac{3}{4}$ in.	14 ft 6 in.	$-2\frac{1}{4}$ in.
6	14 ft $3\frac{1}{8}$ in.	14 ft 6 in.	$-2\frac{7}{8}$ in.
7	14 ft $3\frac{3}{8}$ in.	14 ft 6 in.	$-2\frac{5}{8}$ in.
8	14 ft $3\frac{7}{8}$ in.	14 ft 6 in.	$-2\frac{1}{8}$ in.
9	14 ft $3\frac{5}{16}$ in.	14 ft 6 in.	$-2\frac{11}{16}$ in.
10	14 ft $3\frac{5}{8}$ in.	14 ft 6 in.	$-2\frac{3}{8}$ in.

cision will be defined. Precision is concerned with the repeatability of a measurement. Therefore, precision has meaning only when more than one measurement is made. If you were given the measurements of Table 5.1 and did not know the true length, what would you do to the data to determine the measured length? The first point of interest is that individual 3 must have made a mistake in counting the number of times he placed the ruler end to end. This type of error in recording will be called gross mistake. Gross mistakes will be discussed in more detail later in this chapter, but for the present let us assume that this reading is rectified by individual 3 to become 14 ft 3 in. Now, one logical method of determining the measured length is to find the average value (or arithmetic mean) of all readings:

$$\text{average value} = \frac{\sum_{i=1}^{n} (\text{individual values})_i}{n}$$

For the present example (with individual 3 corrected),

$$\text{average value} = \frac{\sum_{i=1}^{10} (\text{individual values})_i}{10} = 14 \text{ ft } 3\frac{1}{2} \text{ in.}$$

Then, precision is defined to be the accuracy of the readings when compared to the average value of the readings:

precision (upper limit) = maximum value − average value
precision (lower limit) = minimum value − average value
precision = maximum of [precision (upper limit)] or
[−precision (lower limit)]

And by our agreement, the measured value is the average value from all of the readings. Therefore,

$$\text{average measured value} = \text{reading taken} \pm \text{precision}$$

For the example of Table 5.1, precision $= \pm\frac{1}{2}$ in. (reading 3 corrected).

If reading 3 had been rectified, the accuracy would be ±3 in. A measurement can be relatively precise and still not be accurate. In a real measurement problem (where the true value is unknown) precision is a very important property, since accuracy cannot be established. A high-precision measurement system has very small numbers for the precision defined here. Both precision and accuracy are often expressed as percentages of the reading taken or percentages of the full-scale reading. For example,

$$\text{per cent accuracy} = \frac{\text{accuracy}}{\text{true value}} \times 100$$

$$\text{per cent accuracy (full-scale)} = \frac{\text{accuracy}}{\text{full-scale value}} \times 100$$

where the full-scale value is the maximum value that can be measured by the measurement system with that scale setting. Replacing the word accuracy with the word precision in these equations gives the per cent precision. Most instruments have a value for per cent accuracy given in the instrument specifications. The distinction between per cent accuracy based on the reading and based on full-scale reading should be kept in mind when purchasing equipment.

A bias-type error exists when there is a consistant (and relatively constant) difference between the measurement-system output and the true value. It has been established in Chapter 3 on sensors that sensors are selected to be relatively insensitive to all inputs other than the desired input. The same care is exercised in the selection of modifying elements and readout devices. However, some of these effects are usually not eliminated. The ruler used to determine the data of Table 5.1 could have been $\frac{1}{6}$ in. too long. If $\frac{1}{6}$ in. were cut off this ruler, the average valued measured by each individual would be larger by the amount $15(\frac{1}{6}$ in$) = 2\frac{1}{2}$ in. This is a systematically additive process and is called a bias error. The same type of error could be introduced by the systematic method used to mark and move the ruler. However, the use of 10 different individuals tends to reduce the possiblility of the method of placing down the ruler in introducing a bias error.

Remember that environmental conditions are major contributors to bias-type errors. Temperature is a variable that affects most sensors (the ruler becomes longer with increase in temperature). Humidity, atmospheric pressure, radiation from all light sources, small vibrations of the building, stray electrical voltages, power-line surges, and many other sources are present in most measurement laboratories. One typical approach to this problem is to control all of the sources of error that introduce a measurable error in the output signal. This means that if these variables do affect the output signal, they will always affect it the same way and a simple calibration can be used to eliminate the resulting biasing error.

All of the variables that make very small contributions to the output signal may not be controlled. Two possible methods of using data measured under these conditions are available. If the output varies from higher values to lower values (compared to true value) in an unpredictable manner, the contributions to output signal might be random in nature. Usually, electrical noise in the air is of this type, although it might also have a biasing voltage impressed. If the output is both high and low with equal probability that any individual reading could be either high or low, the resulting error is random and can be collected with the precision error, since it has the same characteristics. Should this collection result in a precision that is unacceptable, steps must be taken to control the additional variables of the problem. A second method of using data measured where all variables are not controlled is available when the output is consistently either high or low. This type of error can be analyzed, because it is only a bias error.

The data of Table 5.1 can be treated in a very convenient form by correcting the bias error. When the true value is known (either given or determined from calibration with known signals), the bias error is given by

$$\text{bias error} = \text{average measured value} - \text{true value}$$

Then the true value is

$$\text{true value} = \text{average measured value} - \text{bias error}$$
$$= (\text{measured value} - \text{bias error}) \pm \text{precision}$$

For the data of Table 5.1,

$$\text{bias error} = 14 \text{ ft } 3\tfrac{1}{2} \text{ in.} - 14 \text{ ft } 6 \text{ in.} = -2\tfrac{1}{2} \text{ in.}$$

One basic assumption in the treatment of measured data is that the measurements have small random errors or have constant errors of the bias type. If neither of these conditions are met, the basic operation of the measurement system must be considered, and meaningful information can be obtained only after the source for extraneous inputs has been evaluated. A statistical analysis can be worthwhile only after the measured output has been corrected to a form containing a randomly distributed error, with possibly a bias error.

When a true value of a parameter being measured is not available, there is an uncertainty associated with the measured output. This uncertainty is closely associated with the precision error previously discussed. A single reading of the output of a measurement system can be used to predict the value of the measured parameter within a certain limit, but it cannot be used to predict the exact average measured value or to restrict the error bound. It is not always desirable to require many readings of a measured quantity, but it is important to be able to place confidence limits on a limited number of readings. Ideally, we would like to take one reading and be able to use it to predict, as precisely as possible, the true value of the parameter being measured. The degree of precision depends on the uncertainty surrounding the measurement system and the environment. The most acceptable method of placing values on uncertainty is to provide known inputs (standards) to the measurement system, to observe the output readings, and to apply a statistical analysis to the data collected. It might be pointed out that the above process is calibration, and it does provide a method for determining information about accuracy, precision, bias, and uncertainty. The use of "information about" was intentional, to emphasize that in all statistical work there are no absolute values. All information is based on a finite sample space, and only an infinite sample space could contain all possibilities.

This discussion leads directly to the question of confidence in a measured value associated with precision, accuracy, bias, and true value. If you really do not know the value to represent each of the above terms, what is the use of analysis? This is a legitimate and very appropriate question. To borrow money you must have some collateral. For the measurements to be important, you must be prepared to risk your reputation. In simpler terms, you must be willing to bet with certain odds that measurements will be within the precision or accuracy stated. How do you arrive at good odds for your bet? Obviously, you would be more confident to bet if you had made an extremely large number of tests. However, this is usually very impractical if not impossible. Therefore, you must be able to predict the required odds based on a limited amount of data. These mathematical tools must be developed, since you will stake your reputation on their results.

Before you would be willing to bet on any measured value, you would be sure that all gross mistakes had been eliminated. The reading of a pointer on a dial must have the associated scale factor, the counting of the number of times a ruler is laid end to end must be correct, the use of the proper measurement system must be assured, and the proper consideration of all extraneous inputs must be made before any statistical analysis can provide the desired information on odds. These and other considerations are easily stated, but they are not always obvious. Suppose an individual built a measurement system to check the speed of light parallel and perpendicular to the motion of the earth (the Michelson–Morley experiment). He consistently found results that did not agree with those of all other investigators. He asked his contemporaries to check his apparatus. They found agreement with other results. The final analysis demonstrated that his height (approximately 5 ft) introduced parallax into the apparatus and always caused a difference in his measured speed of light. An extremely competent individual with adequate measurement equipment could still introduce error by his data-taking methods. It is not obvious that gross mistakes have been made, but they must be eliminated before a statistical analysis is made.

If we assume that gross mistakes have been eliminated, the next question is how to place odds on the accuracy of measured data. To consider multiple data of a single parameter value (the example of Table 5.1 has only one value, wall length), it is convenient to plot a curve of frequency of occurance versus a measured value. This type of plot is called both a bar graph and a histogram. Figure 5.1 is a presentation of the bar graph of Table 5.1 (with

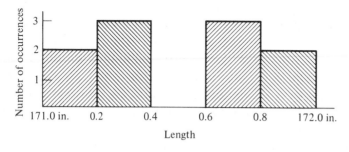

Figure 5.1 Bar Graph of Data of Table 5.1

reading 3 corrected). This type of graph is useful in interpreting the data collected. The measured values are distributed between 171 in. and 172 in. Remember that this data has eliminated the counting error of individual 3 and that this data was hypothetically introduced to illustrate previous definitions. The value measured always depends on the scale divisions of the measurement system, and the number of significant figures available will influence the final analysis.

Based on the limited data of Table 5.1 (with gross mistakes eliminated), you would probably be willing to bet that the value of length measured using the given ruler would be 171.5 in. $\pm$ 0.5 in. in at least 9 out of 10 measurements. That is, you would be willing to bet \$9 to \$1 that the measurement of wall length would be in the range 171.5 ± 0.5 in. if the same measurement system were used. This would be a good bet in the sense that no single reading would be outside this range in the 10 samples. By Las Vegas standards this would almost always pay off for you. To obtain a more secure bet, you must have more individual measurements or expand your precision limits. However, one goal is to obtain the best and most realistic value of precision for your data.

The bar graph of Figure 5.1 may not be the graph that would normally result from a group of independent measurements. We expect that the theories of probability will exert themselves if a large sample space is available. For example, we would expect that 600 throws of a true die would result in a plot similar to Figure 5.2.

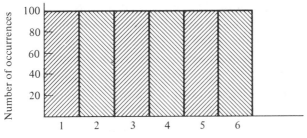

Figure 5.2 Bar Graph of 600 Throws of a Single Die

Numbers occurring

You would also expect the total scores of the toss of two dice to result in a bar graph similar to Figure 5.3, if the dice were thrown a total of 3600 times. Since the probability of throwing 2 is $(\frac{1}{6})(\frac{1}{6}) = \frac{1}{36}$, the probability of throwing 3 is $(\frac{2}{6})(\frac{2}{6}) = \frac{4}{36}$ and so on.

This is not an advertisement to try your luck at the Las Vegas tables. The true percentages of the game are against you. It is an attempt to present realistically the utility of the bar graph.

In measurement there is one average measured value. If there are no extraneous inputs, you should expect a bar graph similar to that of Figure 5.3. There the expected average value would correspond to the number 7 in this figure. Here you would be willing to bet that the total would be

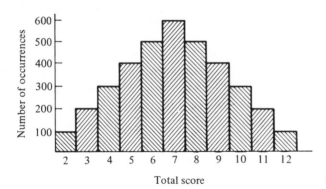

Figure 5.3 Bar Graph of Dice Total in 3600 Throws

7 ± 5 with any odds, say 1 billion to one. Now we have a foundation for comparing a measurement on a statistical basis. For this example there are three ways in which a deviation from the expected graph of Figure 5.3 could occur. First, the dice may not be true dice. That is, the probability of any one side, say 5, coming up might be something other than $\frac{1}{6}$. This is identical to the measurement system that does not have a random distribution about the average value. The randomness of the extraneous inputs must be assured before the measurement system can be analyzed. Second, one of the dice might have numbers from 2 to 7 instead of from 1 to 6. This would add one to the total score of the abscissa in Figure 5.3, otherwise the curve would be the same. This is a bias-type error; the curve will most likely be shifted in the measured value. Finally, with a limited amount of data (3600 is a relatively large sample space) it is possible for the data to be distributed similar to that of Figure 5.4 for 24 throws. It would still be profitable to bet that the total would be 7 ± 5. We are looking for more information about the phenomena being represented. Recall that the bar graph of the data of Table 5.1 was more like Figure 5.4 than Figure 5.3. Two major questions need to be discussed at this point:

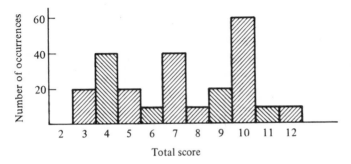

Figure 5.4 Bar Graph of Dice Total in 24 Throws

1. How much data must be collected before probability limits can be set?
2. How can statistical tools be used to analyze the data collected from a measurement system?

The answers to these questions are inherent in the basic formulation of statistical analysis.

We have looked at three types of distributions of data in Figures 5.2, 5.3, and 5.4. In Figure 5.2 the events are equally likely. In Figure 5.3 the events are orderly arranged around the most likely total of 7, and in Figure 5.4 the events are somewhat disorganized. If a measurement system has been used to measure a constant value of the variable, the most likely distribution that would result from many sample readings would be that of Figure 5.3, where the number 7 would represent the average measured value. This ideal distribution can be realized if the extraneous inputs are randomly distributed around the average value. (Calibration would correct data so that the average value would equal the true value.) Therefore, if the measurement system has been properly designed and if the control variables correct for extraneous inputs other than randomly distributed ones, a distribution of measured value should look like Figure 5.3. This curve is a finite (finite number of data points) random distribution.

A mathematical model of Figure 5.3 has been developed for the case of an infinite sample space. The model is used extensively in statistical analysis and takes the form of an integral probability curve. Some feeling for the curve can be obtained by developing probability concepts from the bar graph of Figure 5.3. The probability that on a single throw of the dice the dice total will equal 2 is $\frac{1}{36}$. If the number of occurrences on Figure 5.3 is divided by the total number of tosses, it will be seen that the ordinate actually becomes the probability (for example, the 600 occurrances of the number 7 divided by 3600 gives the probability of throwing a 7 on a single toss, $\frac{1}{6}$). This process of dividing the ordinate by the area under the bar graph (this takes the absissa, 2, 3, 12, as having unit length) is called normalizing the bar graph. Then the probability that a number between 2 and 10 will be thrown is the sum of the area under the normalized curve from 2 to 10. In this case it is $\frac{1}{36} + \frac{2}{36} + \frac{3}{36} + \frac{4}{36} + \frac{5}{36} + \frac{6}{36} + \frac{5}{36} + \frac{4}{36} + \frac{3}{36} = \frac{11}{12}$. Eleven times out of 12 tosses a number between 2 and 10 should show. This analysis is still for a discrete event. In a measurement system data is more continuous, although the number of significant figures and the scale divisions do, in fact, give discrete data. The molecules in a thin gold foil sheet look continuous from 10 ft, but high-energy radiation sources see it as an ordered discrete arrangement. In mathematical statistics we take enough data and stand far enough away that it appears to be a continuous representation. The mathematical model can then be described to encompass the concepts introduced in a measurement system.

The Gaussian or normal distribution has given the most realistic model for a properly analyzed measurement system of all proposed models. *This model has the characteristic that any errors present must be normally distributed around the mean value or the model will not provide a good fit.* Since this is also the goal of a measurement system, conceptually the model should be adequate. Now there remains a presentation of the model and a method to check the reliability of the model in representing the phenomenon.

An integral-type probability function is introduced so that the probability that a measurement is between negative and positive infinity is one,

$$P(-\infty < x < \infty) = \int_{-\infty}^{\infty} f(x)\,dx = 1 \qquad (5.1)$$

This is the normalizing property of the probability function. The function $f(x)$ should be such that it is symmetric about the average measured value (arithmetic mean). It should also be flexible in allowing for various degrees of precision in a measurement system. One particular function can satisfy these criteria,

$$f(x) = \frac{1}{\sigma\sqrt{2\pi}}\,e^{(x-x_m)/2\sigma^2} \qquad (5.2)$$

where x_m is the average or arithmetic mean value of the reading and σ^2 has the role of the precision. The coefficient $1/\sigma\sqrt{2\pi}$ is the constant required to normalize the integral function.

Although there are many functions capable of satisfying Equation 5.1, the function of Equation 5.2 has many satisfying features. It can be presented in a simple tabular form, it has a mean value, it is an even function around this mean value, and it provides flexibility in precision. This particular function is called the Gaussian or normal distribution function. Remember that this is a mathematical model proposed to satisfy the physical situation that exists when measurement systems are used.

We must learn enough about the Gaussian distribution to be able to represent measured data in this form. This will be accomplished by a careful presentation of the graphical and tabular forms of the Gaussian distribution. We then want to have a method of testing to see whether or not the proposed normal distribution does represent the physical problem being considered (the measured data). Finally, we want to be able to interpret the probability function and the tests to provide the odds for obtaining the average measured value with a given precision based on a single measurement. This last sentence is a statement of the product desired from the field of statistical analysis.

THE GAUSSIAN DISTRIBUTION

A plot of the function of Equation 5.2 is given in Figure 5.5 along with the probability function of Equation 5.1 with limits on the integral

$$P(-\infty < x - x_m < x_1) = \int_{-\infty}^{x_1} f(x)\, dx \tag{5.3}$$

This gives the probability that $x - x_m$ is less than x_1.

When a measurement system gives an output, the precision of the system dictates that the average measured value is the measured value plus or

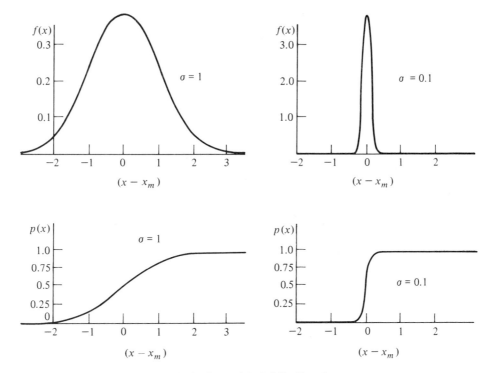

Figure 5.5 Gaussian Distribution and Probability Function

minus the precision. The measurement statement for data similar to that of Figure 5.5 would be

$$\text{predicted value} = x_m \pm \Delta x$$

where Δx is the precision. Now, if the bar graph of measured data could be represented by curves similar to those of Figure 5.5, the question most commonly to be answered is: What is the probability that the true value of the parameter lies between $x_m - \Delta x$ and $x_m + \Delta x$? (If the system has been calibrated and all extraneous inputs controlled with the exception of precision-type inputs, the predicted value could be used to bound the true value of the parameter.) The answer is shown graphically in Figure 5.6.

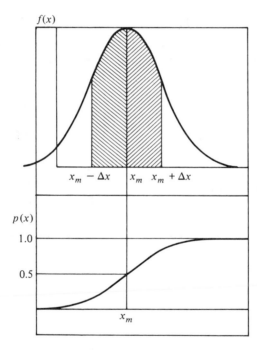

Figure 5.6 Distribution Curve with Precision Type Limits

The probability could be determined using Equation 5.3,

$$P(-\infty < x < x_1) = P(-\infty < x < x_m + \Delta x) = \int_{-\infty}^{x_m + \Delta x} f(x)\, dx$$

$$P(-\infty < x < x_m - \Delta x) = \int_{-\infty}^{x_m - \Delta x} f(x)\, dx$$

Therefore,

$$P(x_m - \Delta x < x < x_m + \Delta x) = \int_{-\infty}^{x_m + \Delta x} f(x)\, dx - \int_{-\infty}^{x_m - \Delta x} f(x)\, dx$$

(5.4)

Equation 5.4 is useful when tabulated values for the integrals are used. Since the function $f(x)$ is very difficult to handle mathematically, values of the probability function, $P(x)$, are published in various sources. The bibliography at the end of this chapter can be used for these sources and for more details in the presentation. A shorter table is presented here in Table 5.2. One table can be valid for all values of x_m and σ by first always shifting the x axis to the point where x_m is located at zero on this axis. This will change Equation 5.4 to

$$P(-\Delta x < x < \Delta x) = \frac{1}{\sigma\sqrt{2\pi}} \int_{-\Delta x}^{\Delta x} e^{-x^2/2\sigma^2}\, dx$$

TABLE 5.2[a]   **Probability Function $P(\alpha x) = P(-\alpha x < v < \alpha x)$ for Gaussian Distribution**

αx	$P(\alpha x)$	αx	$P(\alpha x)$	αx	$P(\alpha x)$
0.00	0.000	0.55	0.563	1.2	0.910
0.05	0.056	0.60	0.604	1.3	0.934
0.10	0.113	0.65	0.642	1.4	0.952
0.15	0.168	0.70	0.670	1.5	0.966
0.20	0.223	0.707	0.682	2.0	0.995
0.25	0.276	0.75	0.711	1.00	
0.30	0.329	0.80	0.742		
0.35	0.379	0.85	0.771		
0.40	0.428	0.90	0.797		
0.45	0.476	0.95	0.821		
0.487	0.500	1.00	0.843		
0.50	0.521	1.1	0.880		

[a] Table is reproduced from *Theories of Engineering Experimentation*, 2nd ed., by H. Schenck, Jr. (Reference 6, p. 21).

Then, let $\sigma = 1/\alpha\sqrt{2}$ and use this change of variables to obtain

$$P(-\Delta x < x < \Delta x) = \frac{1}{\sqrt{\pi}} \int_{-\Delta x}^{\Delta x} e^{-(\alpha x)^2}\, d(\alpha x)$$

or the probability that the variable v lies between $-\alpha x$ and $+\alpha x$ of zero is

$$P(\alpha x) \equiv P(-\alpha x < v < ax) = \frac{1}{\sqrt{\pi}} \int_{-\alpha x}^{\alpha x} e^{-(\alpha x)^2} \, d(\alpha x)$$

If α could be determined for a given measurement, it would be possible to give the probability that a single measurement could be used to obtain the true values within precision limits. Again, it must be emphasized that the data must be distributed in Gaussian form before this probability function can be valid. The limitations of any mathematical model must be understood and respected. As an example of Table 5.2, with $\alpha = 1(\sigma = 1/\sqrt{2})$, 995 times out of 1000 the value measured will be correct ± 2 units of measure. With $\alpha = 10(\sigma = 1/10\sqrt{2})$, 995 times out of 1000 the value measured will be correct ± 0.2 units of measure. In both examples 2 and 0.2 represent the precision of the measurement, and when σ became smaller the precision has a smaller value. *It can be seen that precision is directly related to σ for each value of the probability function.* This gives a feel for σ, but a mathematical expression must be developed to facilitate data evaluation.

The average value (or arithmetic mean value) of a set of data has already been defined:

$$x_m = \frac{1}{n} \sum_{i=1}^{i=n} x_i \tag{5.5}$$

Precision is concerned with the variation of any individual reading from this mean value. Let us define this deviation to be

$$d_i = x_i - x_m \tag{5.6}$$

At first glance it might be thought that an average deviation could be the parameter required, but from Equation 5.5,

$$\text{average deviation} = \frac{1}{N} \sum_{i=1}^{i=N} (x_i - x_m) = \frac{1}{N} \sum_{i=1}^{i=N} x_i - x_m$$

$$= \frac{1}{N} \sum_{i=1}^{i=N} x_i - \frac{1}{N} \sum_{x-1}^{i=N} x_i = 0$$

That is, the average deviation is zero. A meaningful evaluation of the deviation results from applying the root-mean-square technique to deviation data, and σ is defined as

$$\sigma = \left[\frac{1}{N} \sum_{i=1}^{i=N} (x_i - x_m)^2 \right]^{1/2} \tag{5.7}$$

Sigma (σ) is the root-mean-square deviation. It would also be possible to define a deviation using the absolute-value technique to eliminate algebraic cancellation of deviation:

$$\beta = \frac{1}{N} \sum_{i=1}^{i=N} |d_i|$$

However, this alternative method is not to be used with Table 5.2. The standard deviation, σ, is defined by Equation 5.7. The standard deviation squared, σ^2, is a statistical definition of variance. These terms are applied in the statistical analysis of data.

If the data collected in the measurement of wall length (Table 5.1) is analyzed according to the previous statistical presentation, the standard deviation is approximately 0.314. Therefore, α (from Table 5.2) is approximately 2.25. The precision from that data was found to be ±0.5 in. From Table 5.2, the value of $\alpha x = 1.125$ gives the probability using linear interpolation that 887 times out of 1000 measurements the measurement will be 14 ft 3.5 in. $\pm$ 0.5 in. If you had made the initial bet proposed ($10 to $1), you would have collected $887 and paid out $1130. Under these conditions your initial bet would have been a poor choice. However, there is still the question about testing data to determine whether or not it can be represented by the Gaussian distribution model.

Engineers are in the business of betting their reputations on the odds they determine for a given analysis. To obtain a clearer picture of the possibilities of errors, *two tests have been developed to help determine whether or not the data collected can be represented by the Gaussian distribution.* Also, when the true value of the parameter being measured is unknown, a more conservative and statistically more correct value for the standard deviation is determined from

$$\sigma_u = \left[\frac{1}{N-1} \sum_{N=1}^{i=N} (d_i)^2 \right]^{1/2} \tag{5.8}$$

If Equation 5.8 is used, $\sigma = 0.331$, $\alpha = 2.14$, $x = \pm0.5$, and 869 times out of 1000 the measurement will be 14ft 3.5 in. $\pm$ 0.5 in. Here you would collect $869 and pay out $1310. Equation 5.8 is used when the best value (or average value) is not known before the data is collected. This is frequently the situation in measurement. For the measurement of wall length the best value was unknown before data was collected for the given measurement system. Therefore, the probability was $\frac{869}{1000}$. If you wished to place the odds for betting on the measured lengths, a bet of $\frac{869}{131}$ to 1 ($6.63 to $1.00) would make a net profit of $0.47 in 1000 tries.

TESTS FOR NORMAL DATA FIT

Two statistical tests have been devised to evaluate the mathematical model (Gaussian distribution) proposed for application to measured data. The only way to be absolutely certain that data does fit the Gaussian model is to collect an infinite amount of data. All real problems are evaluated in terms of the probability that the data are normally distributed. Although the data may not be exactly normally distributed, the assumption that they are may be valid. For example, the dial on a pressure gage has stops at 0 and 200 psig. Readings outside this range cannot be taken. Therefore, there can be no values of the pressure at negative infinity or at positive infinity. Still, the approximation of a normal distribution might be very appropriate. Figure 5.7 is a comparison of data collected and the normal distribution.

The test for significance is a critical step in data analysis, because it can give strong indications when data is not normally distributed. Suppose that a test of data resulted in a prediction that there was only one chance in 100 that the data was normally distributed. This would certainly cause

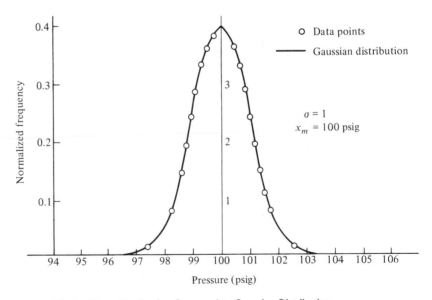

Figure 5.7 Real Data Distribution Compared to Gaussian Distribution

serious doubt about the experimental program. Many sources of error are present in any program, and a danger signal is the start of a search to eliminate them. In a measurement system, one would immediately inspect the system for malfunction and look for sources of extraneous inputs. This would be especially true for calibration data. Other measurement data could have errors resulting from the basic formulation of the problem.

The first test to be considered is the chi squared test for significance. The fundamental development of both tests to be considered here can be found in most texts on statistical analysis. Only the information required to apply these tests will be presented here. Chi squared, χ^2, is defined to be

$$\chi^2 = \sum_{i=1}^{i=N} \left\{ \frac{[(\text{observed events in a range}) - (\text{expected events in a range})]^2}{(\text{expected events in a range})} \right\}$$

$$i\text{th range} \qquad (5.9)$$

In comparing observed data in a given range to the data that would be obtained in that range if the data were normally distributed, the expected value must be obtained for the mean value and a standard deviation calculated from the measured data. *A bar graph of the observed data should be plotted and ranges of the observed value should be selected in such a way that the frequency of occurrence in each range is approximately equal.* There must be at least four range divisions made, but actually more ranges are desirable. (For four division points, x_1, x_2, x_3, x_4, the ranges of observed values would be $-\infty$ to x_1, x_1 to x_2, x_2 to x_3, x_3 to x_4, and x_4 to $+\infty$. This would give five ranges, with the points selected to ensure that no observation has that value.) *It is also desirable to have at least four values of the observed data in each range.* For five ranges this would require at least 20 observations. The probability function for the normal distributation can be determined from Table 5.2. The expected number of events in a given range would then be the probability for that range multiplied by the total number of observations.

Once chi squared has been determined and the number of ranges known, Figure 5.8 can be used to determine a probability number. *The number of degrees of freedom is equal to the number of ranges selected minus 3.* Since zero degrees of freedom is not meaningful, the reason for requiring at least four ranges is obvious. It is also apparent that the greater the number of degrees of freedom the greater will be the restrictions on chi squared.

If the observed data did, in fact, result from a Gaussian distribution, there is one chance in 20 that chi squared could be larger than the $P = 0.05$ curve for the given degrees of freedom. There are also 19 chances out of 20

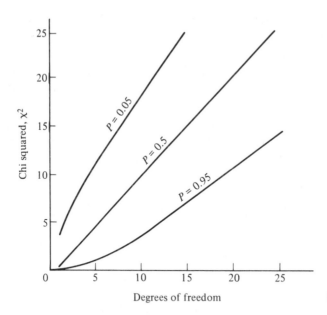

Figure 5.8 Chi-Squared Test Curve*

that normally distributed data will give a chi-squared value that lies above the $P = 0.95$ curve for the given degrees of freedom. Remember that there is no absolute assurance to be gained from this test. However, when the number of observations increase, more confidence can be placed on the test.

The selection of the 5 per cent and the 95 per cent lines was made primarily to provide a dividing region for discussion. Since there is no absolute dividing line between data that are normally distributed and those that are not, these lines also serve to caution the user that there is some doubt that the data collected was actually normally distributed. A rule of thumb is to review the data source carefully when chi squared is larger than the value on the $P = 0.05$ curve or smaller than the $P = 0.95$ curve. In the former case, when chi squared is larger than the value on the $P = 0.05$ curve, extraneous inputs and uncontrolled parameters may be present in the system. In the latter case the measurement system may not be responding to the input signal, or it may not be sensitive enough to function properly.

Although this test cannot give complete assurance, it does provide a checkpoint for evaluating data. At this point the assumption that the data was normally distributed will be shown to be reasonable ($P < 0.8, P > 0.2$), questionable ($P > 0.95$, $P < 0.05$), or doubtful ($P > 0.99$, $P < 0.01$). It should again be emphasized that if the probability is only 0.0001, the data

* Figure is reproduced from *Experimental Methods for Engineers*, 1st. ed., by J. P. Holman (Reference 3, p. 57).

could still be a product of a normally distributed phenomenon (one time in 10,000 tries).

EXAMPLE 5.1

The data of Table 5.1 contain only 10 observations and do not satisfy the desired requirements of the chi-squared test. Let us suppose, therefore, that each individual made two measurements and that they were identical. With the 20 observations (171, 171, $171\frac{1}{8}$, $171\frac{1}{8}$, $171\frac{1}{4}$, $171\frac{1}{4}$, $171\frac{5}{16}$, $171\frac{5}{16}$, $171\frac{3}{8}$, $171\frac{3}{8}$, $171\frac{5}{8}$, $171\frac{5}{8}$, $171\frac{11}{16}$, $171\frac{11}{16}$, $171\frac{3}{4}$, $171\frac{3}{4}$, $171\frac{7}{8}$, $171\frac{7}{8}$, 172, 172) it would be possible to follow the procedure of the previous discussion. We select five ranges, requiring that four observations occur in each range. For this data

$$x_m = 171.5$$

and

$$\sigma_u = \sqrt{\frac{1}{N-1} \sum_{i=1}^{i=N} (d_i)^2}$$

$$\alpha \doteq 2.171$$

Now the ranges are arbitrarily selected to determine the probability from Table 5.2, the fact that the Gaussian distribution is symmetric about the mean value and the concepts of probability are used.

Range	Probability From Table 5.2	Expected Events	Observed Events	χ_i^2
$-\infty$–171.2	0.1785	3.57	4	0.052
171.2–171.35	0.1442	2.88	4	0.433
171.35–171.65	0.3546	7.092	4	1.348
171.65–171.8	0.1442	2.88	4	0.433
171.8–$+\infty$	0.1785	3.57	4	0.052

$$\chi^2 = 2.318$$

Figure 5.9 shows the Gaussian distribution; the shaded portion represents the probability function of Table 5.2. From this presentation

$$P(-\infty < v < x_m - \alpha x) = \frac{1 - P(\alpha x)}{2}$$

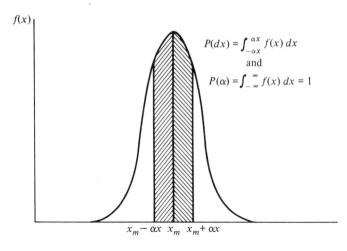

$$P(dx) = \int_{-\alpha x}^{\alpha x} f(x)\, dx$$

and

$$P(\alpha) = \int_{-\infty}^{\infty} f(x)\, dx = 1$$

Figure 5.9 Gaussian Probability Function

and

$$P(-\infty < v < x_m + \alpha x) = P(\alpha x) + \frac{1 - P(\alpha x)}{2} = \frac{1 + P(\alpha x)}{2}$$

and

$$P(x_1 < v < x_2) = P(-\infty < v < x_2) - P(-\infty < v < x_1)$$

with

$$x_2 > x_1$$

where $P(-\infty < v < x_m - \alpha x)$ is the probability that a value, v, would be observed in the range between $-\infty$ and $x_m - \alpha x$. This is also the area under the Gaussian distributation between these limits.

Now, with the degrees of freedom equal to 2 (5 ranges − 3) and chi squared equal to 2.318, Figure 5.8 gives the probability as approximately 0.3. Therefore, there is no evidence that the data are not normally distributed. (This result does not prove that the data are normally distributed!)

EXAMPLE 5.2

The outside temperature was measured 10 times with each of two different measurement systems. The results were 79, 80, 80, 82, 78, 79, 80, 82, 79, 80 with system I; 85, 88, 84, 88, 85, 88, 85, 84, 86, 88 with system II. Use the chi-squared test for significance for the data below:

Ranges	Probability From Table 5.2	Expected Events	Observed Events	χ_i^2
$-\infty$–79.0001	0.116	2.31	4	1.236
79.0001–81.9999	0.262	5.24	4	0.293
81.9999–84.0001	0.244	4.88	4	0.159
84.0001–87	0.262	5.24	4	0.293
87–$+\infty$	0.116	2.31	4	1.236

$$\chi^2 = 3.217$$

For this value of chi squared and two degrees of freedom, $P = 0.10$. Again the data could have been from a normally distributed phenomenon.

For a small sample space, which is common in measurement systems, the chi-squared test is not as sensitive to the data as is desirable. However, this test cannot interpret data, and the results of the data of Example 5.2 indicate a problem in the calibration of the measurement systems or some extraneous input (only one sensor is in the sun, wind is blowing on only one sensor, improper scale reading, etc). The field of statistics was not developed to eliminate the responsibility of the engineer to think, but it is a powerful tool in the hands of competent persons.

A second test was developed for the consideration of data from a small sample. Distribution tables have been formed for data collection less than the infinite sample required for the Gaussian distribution. The mathematical development of the statistics of small sample spaces was made by W. G. Gossett and published under the name Student. Although the distributation tables could be used to determine more realistic precision values for small samples, the development is also a powerful tool for evaluating the probability that two different sample spaces result from the same phenomenon. Other applications are available, but the use discussed here is directed toward comparing two sets of data. For example, if two individuals measured the same phenomenon with the same measurement system, if two measurement systems were used to measure the same phenomenon, or if the same measurement system was used to measure two separate but seemingly identical experiments, the question might arise concerning the validity of assuming that the individuals made precise measurements, that the two measurement systems were equivalent, or that the separate experiments were identical.

The mathematical development of Student's "t" test is capable of answering questions of the type posed in the preceding paragraph. The hypothesis checked in each case is that the two samples of data are identical. The proof is based on the definition

$$t = \frac{x_{m_1} - x_{m_2}}{S\sqrt{1/N_1 + 1/N_2}} \tag{5.10}$$

where the mean values of sample 1 minus the mean value of sample 2 is the numerator,

$$S = \left(\frac{\sum x_1^2 + \sum x_2^2}{N_1 + N_2 - 2}\right)^{1/2} \tag{5.11}$$

where $x_1 = $ (observed of sample 1) $- x_{m_1}$, and N is the number of samples in either sample 1 or sample 2. The number of degrees of freedom is $N_1 + N_2 - 2$. Then Figure 5.10 can be used to determine the probability that both samples are identical. Here again, a probability of less 0.05 would generally indicate that the data should be seriously questioned, since only

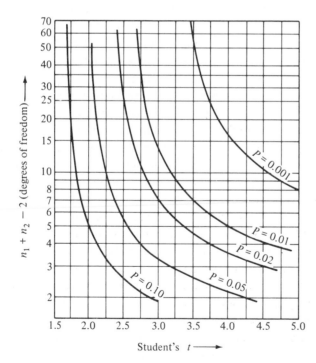

Figure 5.10 Student's "t" Test for Identical Sources*

* Figure is reproduced from *Theories of Engineering Experimentation*, 2nd. ed., by H. Schenck, Jr. (Reference 6, p. 183).

one time in 20 would normally distributed data display this type of representation. The discussion or probability considerations are the same as those applied to the chi-squared test.

EXAMPLE 5.3

If the two samples of data from Example 5.2 are checked using Student's "t" test, is it probable that both measurement systems are equivalent?

Observed 1	x_{m_1}	$(x - x_m)^2$	Observed 2	x_{m_2}	$(x - x_m)^2$
79	79.9	0.81	85	86.1	1.21
80		0.01	88		3.61
78		3.61	84		4.41
79		0.81	88		3.61
80		0.01	85		1.21
82		4.41	88		3.61
80		0.01	85		1.21
82		4.41	84		4.44
79		0.81	86		0.01
80		0.01	88		3.61
Totals		14.90			26.90

$$S = \left(\frac{41.8}{18}\right)^{1/2} = 1.524$$

$$t = \frac{79.9 - 86.1}{1.524\sqrt{\frac{1}{10} + \frac{1}{10}}} = -9.1$$

The minus sign results from the arbitrary selection of which sample is sample 1 and can be neglected with no loss in significance of the test. Therefore, with this value and the degrees of freedom ($10 + 10 - 2 = 18$), the probability that the measurement systems are identical is found, from Figure 5.10, to be much less that 0.001. From this it is very unlikely that the two systems were identical. Again the source may be an extraneous input, or the systems may be uncalibrated.

A more sophisticated statistical analysis of data could be made by a competent statistician, but the sort of results obtained by this type of analysis would be very similar to the sort obtainable from the tests proposed in this book. Most of the problems faced in normal engineering tests can be evaluated using the tools already available. If additional consideration is required, the references listed at the end of this chapter are available for source information. Most measurement problems in data analysis can be resolved using simple mathematical tools. The interpretation of results is clear for the analysis of static measurements with single-value input signals.

When it is not convenient to take 20 measurements of each input signal, some additional statistical methods can be developed for special cases. The general requirements that must be met for these cases are that a known functional relationship exist between input and output and that several (greater than 20 if possible) outputs be measured. For example, if a plot of output versus input is a straight line, 20 readings of output for various inputs could be analyzed statistically. This is also true for higher-order curves, but the calculations are much more lengthy.

LEAST-SQUARES DATA FIT

One mathematical tool that is essential for experimental work is a method of finding the best mathematical expression for the relationship between input and output information. The criterion to be used to define the "best" mathematical fit can be varied for different situations. One accepted standard criteria for the "best fit" is the least-squares method. Let us look at possible "best-fit" criteria for the straight-line problem. Since the applications for least-squares data fit are not limited to a few situations, it seems appropriate that the reader should be capable of developing a least-squares fit to a set of data that cannot be fitted to a straight line. Therefore, the techniques required to accomplish this goal are included in this section. It might be very appropriate for the reader to use the results of this section in the form of Equation 5.16 or Equations 5.21 and 5.22 at this time. The development of these equations and the presentation for an nth-order polynomial are included to assist in situations where the results of this section are not adequate.

The attempt to fit collected data to a straight-line plot involves fitting observed ordered pairs $(x_1, y_1), (x_2, y_2), \ldots, (x_n, y_n)$ to a curve $y = mx + b$, where m and b are to be determined. If only two ordered pairs were available, m and b would be determined exactly from

$$y_1 = mx_1 + b$$
$$y_2 = mx_2 + b$$

This is straightforward, since two points exactly determine a single straight line. (Also, any number, n, of ordered pairs will determine a polynomial of degree $n - 1$ exactly. Three pairs will exactly determine a parabola, four pairs will exactly determine a cubic equation, etc.) Therefore, the first need for a criterion to fit ordered pairs to a curve exists when the number of ordered pairs minus one is greater than the degree of the polynomial being fit. For the straight line, the polynomial of degree 1, at least three ordered pairs must be available before a fitting technique is required.

One possible criterion for a "best fit" is simply to require that the summation of all the distances of each ordered pair from the line be zero. This criterion could give either of the curves of Figure 5.11 for the ordered pairs. Any

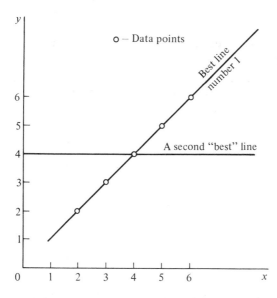

Figure 5.11 Best Fit Criteria for Least Distance From the Line

straight line through the point (4, 4) would satisfy this criterion. This is not a completely general example, since the ordered pairs are on a straight line; but even if they were not, there would be many curves that could satisfy the criteria for any data. The only difference would be the calculation time required to determine these lines.

The least-squares method requires that the distances from the ordered pair to the curve first be squared (to eliminate the algebraic sign) and then summed. The line is then chosen so that this sum has its least value. Remember that the constants, m and b, will be determined by application of a "best-fit" criterion. This will be generalized for higher-degree polynomials later in this section. For the least-squares fit of a straight line, Figure 5.12 is

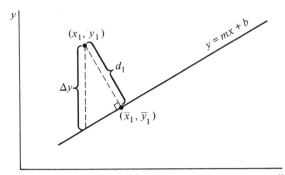

Figure 5.12 Graphical Presentation of Least Squares Analysis

used for the analysis of one ordered pair. To find the distance, d_1, between (x_1, y_1) and the line $y = mx + b$, the intersection of the line normal to $y = mx + b$ must be found. The equation of the normal line can be found by the point–slope formula, since the slope of the normal line is the negative reciprocal of the slope of the given line, $y = mx + b$. Therefore, the slope of the given line is

$$\frac{dy}{dx} = \frac{d(mx + b)}{dx} = m$$

and the slope of the normal line is $-1/m$. The equation of the normal line is

$$y - y_1 = -\frac{1}{m}(x - x_1)$$

The point of intersection of the normal line and the curve can be found by the simultaneous solution of the two equations. This intersection point will be called $(\bar{x}_1, \bar{y}_1)$.

(1) $$y = mx + b$$

(2) $$y = -\frac{x}{m} + \frac{x_1}{m} + y_1$$

Subtracting 2 from 1 gives

$$0 = \left(m + \frac{1}{m}\right)x + b - \frac{x}{m} - y_1$$

or

$$x = \frac{(x_1/m) + y_1 - b}{m + 1/m} = \bar{x}_1$$

and substituting $\bar{x}_1$ into Equation (1) gives

$$y = \frac{x_1 + my_1 + b/m}{m + 1/m} = \bar{y}_1$$

The distance between (x_1, y_1) and $(\bar{x}_1, \bar{y}_1)$ is given by the distance formula

$$d_1^2 = (x_1 - \bar{x}_1)^2 + (y_1 - \bar{y}_1)^2 \tag{5.12}$$

$$= \left(\frac{1}{m + 1/m}\right)^2 \left[(y_1 - mx_1 - b)^2 + \left(x_1 - \frac{y_1}{m} + \frac{b}{m}\right)^2\right]$$

Between two other points (x_2, y_2) and the line, the distance is

$$d_2^2 = \left(\frac{1}{m + 1/m}\right)^2 \left[(y_2 - mx_2 - b)^2 + \left(x_2 - \frac{y_2}{m} + \frac{b}{m}\right)^2\right]$$

Now, the summation of the squares is

$$\sum_{i=1}^{n} d_i^2 = \sum_{i=1}^{n} \left(\frac{1}{m + 1/m}\right)^2 \left[(y_i - mx_i - b)^2 + \left(x_i \frac{y_i}{m} + \frac{b}{m}\right)^2\right] \tag{5.13}$$

The value of $\sum d_i^2$ can be controlled only by the selection of m and b, since all other terms in the expression are fixed by the data of ordered pairs. To find the optimum value of the $\sum d_i^2$, the total differential of $\sum d_i^2$ must be equal to zero. (In this case the optimum value will be a minimum, since the selection of $+\infty$ or $-\infty$ for either m or b will give $+\infty$. Also, the sum is always positive for real values of m and b.) The total differential is set equal to zero.

$$d(\sum d_i^2) = \frac{\partial(\sum d_i^2)}{\partial m} dm + \frac{\partial(\sum d_i^2)}{\partial b} db = 0$$

The total differential can be zero for the general case when the coefficients of dm and db are zero. Or

$$\frac{\partial(\sum d_i^2)}{\partial m} = 0 \quad \text{and} \quad \frac{\partial(\sum d_i^2)}{\partial b} = 0$$

Equation 5.13 used with the above requirements gives

$$\frac{\partial(\sum d_i^2)}{\partial m} = 0 = \sum_{i=1}^{n} \left\{ \left(\frac{1}{m + 1/m} \right)^2 \left[2(y_i - mx_i - b)(-x_i) \right. \right.$$

$$\left. \left. + 2\left(x_i - \frac{y_i}{m} + \frac{b}{m} \right)\left(\frac{y_i}{m^2} - \frac{b}{m^2} \right) \right] + \frac{(2/m^2 - 2)}{m + 1/m} d_i^2 \right\}$$

Multiplying both sides of the above equation by $(m + 1/m)^2/2$ gives

$$0 = \sum_{i=1}^{n} \left\{ \left[(y_i - mx_i - b)(-x_i) + \left(x_i - \frac{y_i}{m} + \frac{b}{m} \right)\left(\frac{y_i}{m^2} - \frac{b}{m^2} \right) \right] \right.$$

$$\left. + \left(\frac{1}{m^2} - 1 \right)\left(m + \frac{1}{m} \right)(d_i^2) \right\}$$

Similarly,

$$\frac{(m + 1/m)^2}{2} \frac{\partial(\sum d_i^2)}{\partial b} = 0 = \sum_{i=1}^{n} \left[(y_i - mx_i - b)(-1) \right.$$

$$\left. + \left(x_i - \frac{y_i}{m} + \frac{b}{m} \right)\left(\frac{1}{m} \right) \right]$$

It is possible to solve the last expression for b in terms of m, where all summations are understood to go from 1 to the number of ordered pairs, n. Also, the summation sign is distributive over the individual additive elements $[\sum(x_i + y_i) = \sum x_i + \sum y_i]$. Then the last expression becomes

$$m \sum x_i + nb - \sum y_i + \frac{1}{m} \sum x_i - \frac{1}{m^2} \sum y_i + \frac{nb}{m^2} = 0$$

$$\frac{nb}{m^2}(m^2 + 1) = \frac{\sum y_i}{m^2}(m^2 + 1) - \frac{m \sum x_i}{m^2}(m^2 + 1)$$

$$b = \frac{\sum y_i - m \sum x_i}{n} \tag{5.16}$$

When this expression is substituted into Equation 5.14 for b, we have a sixth-order polynominal in m to be solved (there is no reduction formula for this order polynomial).

$$m^6 \left[\frac{\sum x_i \sum y_i}{n} \right] - m^5 \left[\frac{\sum x_i \sum y_i}{n} \right] + m^4 \left[\sum x_i^2 + \frac{2 \sum x_i \sum y_i}{n} - \sum y_i^2 \right]$$

$$+ m^3 \left[\sum x_i^2 - \sum y_i^2 + \frac{\sum x_i \sum x_i + \sum y_i \sum y_i}{n} \right]$$

$$+ m \left[\sum x_i^2 - \frac{\sum x_i \sum x_i}{n} - \frac{\sum y_i \sum y_i}{n} \right]$$

$$+ \left[\frac{\sum x_i \sum y_i}{n} - \sum x_i \sum y_i \right] = 0 \tag{5.17}$$

This is a formidable problem to be solved by hand calculations. All six solutions for m must be checked to eliminate the five extraneous roots. Because this development results in a solution that requires considerable computation, a second consideration of the criterion for "best fit" is made. The above development assumes that both of the ordered pairs, x and y, have errors in their measurement. However, in the calibration of measurement systems one of the variables is known accurately (let us say x). If there is no error in one of the variables, a more realistic criterion would be to require that the distance from the point (ordered pair) to the curve ($y = mx + b$) be measured along a line of constant input signal (x in this case). Referring to Figure 5.12, the distance desired is Δy,

$$\Delta y_1 = y_1 - y$$

where $y = mx_1 + b$. For this case the summation of the distances squared is

$$\sum_{i=1}^{n} \Delta y_i^2 = \sum_{i=1}^{n} (y_1 - mx_i - b)^2$$

The optimum values of m and b can be determined in the manner previously discussed:

$$\frac{\partial[\sum(\Delta y_i^2)]}{\partial m} = 0 = 2 \sum_{i=1}^{n} (y_i - mx_i - b)(-x_i)$$

$$\frac{\partial[\sum(\Delta y_i)^2]}{\partial b} = 0 = 2 \sum_{i=1}^{n} (y_i - mx_i - b)(-1)$$

or

$$\sum x_i y_i - m \sum (x_i)^2 - b \sum x_i = 0 \qquad (5.18)$$

$$\sum y_i - m \sum x_i - bn = 0 \qquad (5.19)$$

The last expression can be solved for b in terms of m:

$$b = \frac{\sum y_i - m \sum x_i}{n} \qquad (5.20)$$

This is identical to the case where both variables can have error, but the value for m can be determined by substitution:

$$\sum x_i y_i - \frac{\sum x_i \sum y_i}{n} = m\left(\sum x_i^2 - \frac{\sum x_i \sum x_i}{n}\right)$$

or

$$m = \frac{n \sum x_i y_i - \sum x_i \sum y_i}{n \sum x_i^2 - (\sum x_i)^2} \qquad (5.21)$$

and

$$b = \frac{\sum x_i^2 \sum y_i - \sum x_i \sum x_i y_i}{n \sum x_i^2 - (\sum x_i)^2} \tag{5.22}$$

Equations 5.21 and 5.22 are used extensively in all fields of experimentation. When data can be represented by a straight line and when one variable is accurately known (in the case of calibration), these equations determine the constants for the least-squares straight line. The equations are also used for cases where both variables have errors, but the effect of one variable is neglected. Equations 5.16 and 5.17 should be used for this type of data.

In a measurement-system calibration process, the input signal is accurately known and the output can be related to the input in some manner. If the system is linear, Equations 5.21 and 5.22 can be used. Then, the object of the measurement system is to sense an unknown input signal and to use the calibration equation to ascertain the value of the input signal. For this type of operation a better criterion for data plot is reflected by Equations 5.16 and 5.17. However, it is common practice to use Equations 5.21 and 5.22 for all situations. If a set of ordered pairs is to be fit to an nth-order curve using the least-squares fit, the development would be the same as shown before,

$$y = a_n x^n + a_{n-1} x^{n-1} + \cdots + a_1 x + a_0$$

The criteria for differences based on Δy gives

$$\sum_{i=1}^{m} (\Delta y_i)^2 = \sum_{i=1}^{m} (y_i - a_n x_i^n - a_{n-1} x_i^{n-1} \cdots a_1 x_i - a_0)^2$$

where m is an integer greater than $n + 1$. The requirement that the coefficient of each da_i equal zero gives $n + 1$ equations for the $n + 1$ a's.

$$\frac{\partial[\sum(\Delta y_i)^2]}{\partial a_n} = 0 = -2 \sum (x_i - a_n x_i^n - a_{n-1} x_i^{n-1} \cdots a_1 x_i - a_0)(x_i^n)$$

$$\frac{\partial[\sum(\Delta y_i)^2]}{\partial a_{n-1}} = 0 = -2 \sum (\Delta y_i)(x_i^{n-1})$$

$$\frac{\partial[\sum(\Delta y_i)^2]}{\partial a_1} = 0 = -2 \sum (\Delta y_i)(x_i) \tag{5.23}$$

$$\frac{\partial[\sum(\Delta y_i)^2]}{\partial a_0} = 0 = -2 \sum \Delta y_i$$

To avoid solving $n + 1$ equations for the $n + 1$ a's, it is desirable to find variables, x and y, which have a linear relationship. This is frequently much easier than would first seem evident. For example, the resistance of a thermistor, R, varies with temperature, T, according to the functional relationship

$$R = R_0 e^{A[(1/T) - (1/T_0)]}$$

When a relationship involves the logarithic function, the technique employed to obtain a linear function is to take the log of both sides of the equation:

$$\ln\left(\frac{R}{R_0}\right) = A\left(\frac{1}{T} - \frac{1}{T_0}\right)$$

Let $x = 1/T$ and $y = \ln R$. Then

$$y - \ln R_0 = Ax + \frac{A}{T_0}$$

or $y = Ax + b$, a linear function. The same technique is applicable for any other base logarithm.

For functions of the type $y = Ax^n$, the first step is to take the logarithm of both sides of the equation:

$$\log(y) = n[\log(x)] + \log A$$

Then, with $\log y = w$ and $\log x = u$,

$$w = nu + b$$

A log–log plot would produce a straight line.

The parabolic function, $y = ax^2 + bx + c$, occurs commonly in measurement systems (the thermocouple over limited ranges). This form can be reduced to a previous type by completing the square,

$$\frac{y}{a} = x^2 + bx + \left(\frac{b}{2}\right)^2 - \frac{b^2}{4} + 2$$

or

$$y = a\left(x + \frac{b}{2}\right)^2 + a\left(c - \frac{b^2}{4}\right)$$

Then $y - y_1 = w$, $x - x_1 = u$, and

$$w = au^2$$

This type of plot is not as desirable as the previous forms because the ordered pair, (x_1, y_1), will lie on the final curve. Therefore, this ordered pair is statistically weighted more than other ordered pairs.

In many cases algebraic manipulation of the functional relationships existing between two variables will produce a new set of variables having a linear relationship between them. It would probably be worthwhile to solve Equation 5.23 for the coefficients (a's) when a polynomial-type function exists between the variables. Computers can be used to solve the square matrix that results from these equations. For the parabolic function Equation 5.23 gives

$$a_2 \sum x_i^4 + a_1 \sum x_i^3 + a_0 \sum x_i^2 = \sum x_i^2 y_i$$
$$a_2 \sum x_i^3 + a_1 \sum x_i^2 + a_0 \sum x_i = \sum x_i y_i$$
$$a_2 \sum x_i^2 + a_1 \sum x_i + a_0 n = \sum x_i$$

Where n is the number of ordered pairs (greater than 3), the results come simply from the solution of three equations with three unknowns.

STATISTICAL STATEMENTS FOR LEAST-SQUARES FITS

The final expression for the best data fit to a curve using a least-squares criterion results when the values of all of the coefficients have been determined from the solution of Equation 5.23. For a first-order curve, the data best fit $y = mx + b$, where m and b are determined from Equations 5.21 and 5.22. (These equations will be used throughout this discussion, understanding that the criterion allows variation of one variable only.) The deviation of the data points from this curve is given by the least-squares condition

$$\sigma^2 = \left(\frac{1}{n-1} \right) \sum_{i=1}^{n} (\Delta y_i)^2 = \frac{1}{n-1} \sum_{i=1}^{n} (y_i - mx_i - b)^2 \qquad (5.24)$$

This standard deviation is based on limited data, where the mean value is not known before the test. If 20 or more ordered pairs (x_i, y_i) are measured, the techniques already presented for the chi-squared test may be used. Now, the variables for the bar graph are the Δy terms. The mean value for all data is given by $y_m = mx_m + b$. The types of statistical statements that can be made are identical to those stated previously under the discussion of the chi-squared test.

If the values of y are recorded for increasing x and then for decreasing x, the effect of hysteresis could be sensed by a Student's "t" test of the two groups of data. The mean value of the data for increasing x would be found at some convenient value, x_{m_1}, and the mean value of the data for decreasing x would be found for the same value, x_{m_2}. Then,

$$S = \left[\frac{\sum (\Delta y_{i_1})^2 + \sum (\Delta y_{i_2})^2}{n_1 + n_2 - 2} \right]^{1/2} \qquad (5.25a)$$

and Student's "t" is given by

$$t = \frac{y_{m_1} - y_{m_2}}{S\sqrt{1/n_1 + 1/n_2}} \qquad (5.25)$$

Again the statements concerning the results of the test are the same as before.

For an nth-order polynomial fit the only change in the above discussion is the definition of Δy:

$$\Delta y_i = y_i - a_n x_i^n - a_{n-1} x_i^{n-1} - \cdots - a_1 x_i - a_0$$

All other discussions are the same.

5.4
Data Analysis of Total Problems

The statistical tools available at this point cannot effectively describe transient, or dynamic, phenomena. This discussion will include dynamic systems in following sections. Now the basic tools will be applied to the analysis of problems involving steady-state (static) phenomena. The ultimate goal of any experimental program is to be able to predict the behavior of the dependent variables, based on a knowledge of the independent, or control, variables. Associated with this statement, the basic numerical limitations of the mathematical model must be stated, and the reputation of the scientist is one of the stakes.

A competent scientist will certainly be able to state the limitations that should be placed on the results of his research. Three basic avenues are available for this statement: absolute limits, restricted limits, and statistical limits. Absolute limits are of the type

$$y = f(x) \pm \text{absolute limit}$$

For these limits, the value of y is guaranteed by the engineer never to lie outside the range given. Restricted limits are posed when there is still some uncertainty about the effect of certain variables on the physics of the problem. The most common limits placed on a program are statistical limits. These state that

1. For 1σ limit, $y = f(x) \pm \sigma$, 68 per cent of the y values will lie in this range.
2. For 2σ limits, $y = f(x) \pm 2\sigma$, 95 per cent of the y values will lie in this range.
3. For 3σ limits, $y = f(x) \pm 3\sigma$, 99.7 per cent of the y values will lie in this range.

(A certain soap is guaranteed to be 99.44 per cent pure; it is almost pure $\pm 3\sigma$.) The last statement is almost an absolute limit.

The usual product of an experimental program is a mathematical model of a particular phenomenon of the form

$$y = f(x_1, x_2, x_3, \ldots, x_n)$$

where y is the desired output information and the x_i's are independent variables. The method of determining the values of the independent variables is normally by a measurement system. The question naturally arises about how the uncertainty of measuring the independent variables ultimately affects the product output. One convenient method of describing these effects is to consider the incremental change in the product as a function of the uncertainty of the value of the independent variables.

$$y + \Delta y = f(x_1 + \Delta x_1, x_2 + \Delta x_2, \ldots, x_n + \Delta x_n)$$

then

$$\Delta y = \frac{\partial f}{\partial x_1} \Delta x_1 + \frac{\partial f}{\partial x_2} \Delta x_2 + \cdots + \frac{\partial f}{\partial x_n} \Delta x_n$$

If the uncertainties in the independent variables, Δx_i's, are absolute limits,

$$\Delta y_{abs} = \left| \frac{\partial f}{\partial x_1} \Delta x_1 \right| + \left| \frac{\partial f}{\partial x_2} \Delta x_2 \right| + \cdots + \left| \frac{\partial f}{\partial x_n} \Delta x_n \right| \qquad (5.26)$$

and

$$y = f(x_1, x_2, \ldots, x_n) \pm \Delta y_{abs}$$

If the Δx_i's are given in statistical limits ($\sigma, 2\sigma, 3\sigma$), the limits on Δy are similarly represented. The absolute value signs are required to provide for the situation where all Δx_i's are additive. A less conservative uncertainty results from the root-sum-square method, where

$$\Delta y_{rss} = \left[\left(\frac{\partial f}{\partial x_1} \Delta x_1 \right)^2 + \left(\frac{\partial f}{\partial x_2} \Delta x_2 \right)^2 + \cdots + \left(\frac{\partial f}{\partial x_n} \Delta x_n \right)^2 \right]^{1/2} \qquad (5.27)$$

Since the probability of either a plus or minus error in the Δx_i's is equal,

Equation 5.27 is statistically a more realistic representation of the probable error. Again, the resulting Δy has the meaning of the statistical limits imposed on the individual uncertainties of the Δx_i's (σ, 2σ, 3σ).

This discussion completes the statistical effect of data on the final output of the system. However, many of the functional relationships between output and inputs are product–quotient type functions. For example,

$$y = \frac{x_1 x_2 x_4^2 x_5^{1.4}}{x_3}$$

For this type of representation, it is convenient to express Δy in a percentage form, $\Delta y/y$. For all product forms, the per cent uncertainty is simply expressed as

$$\Delta y = \left| \frac{x_2 x_4^2 x_5^{1.4}}{x_3} \Delta x_1 \right| + \left| \frac{x_1 x_4^2 x_5^{1.4}}{x_3} \Delta x_2 \right| + \left| \frac{2 x_1 x_2 x_4 x_5^{1.4}}{x_3} \Delta x_4 \right|$$

$$+ \left| \frac{1.4 x_1 x_2 x_4^2 x_5^{1.4}}{x_3} \Delta x_5 \right| + \left| \frac{-x_1 x_2 x_4^2 x_5^{1.4}}{x_3^2} \Delta x_3 \right|$$

$$\frac{\Delta y}{y} = \left| \frac{\Delta x_1}{x_1} \right| + \left| \frac{\Delta x_2}{x_2} \right| + \left| \frac{2 \Delta x_4}{x_4} \right| + \left| \frac{1.4 \Delta x_5}{x_5} \right| + \left| \frac{-\Delta x_3}{x_3} \right|$$

or, in alternative form,

$$\frac{\Delta y}{y} = \sqrt{\left(\frac{\Delta x_1}{x_1} \right)^2 + \left(\frac{\Delta x_2}{x_2} \right)^2 + \left(\frac{2 \Delta x_4}{x_4} \right)^2 + \left(\frac{1.4 \Delta x_5}{x_5} \right)^2 + \left(\frac{\Delta x_3}{x_3} \right)^2}$$

Therefore, the per cent error, $100 \, \Delta y/y$, is expressed in terms of the per cent error of each individual measurement.

EXAMPLE 5.4

Suppose that the problem is to determine the stress in a structural member by measuring the load applied and the dimensions. If the error in measuring length is ± 0.01 in. (0.0254 cm) and the error in measuring force is ± 1 lb$_f$ (4.448 N) within 2σ limits, what is the error in determining stress?

$$S = \frac{F}{A}$$

For a circular cross section,

$$S = \frac{4F}{\pi d^2}$$

then

$$\Delta S = \left| \frac{\partial (S)}{\partial F} \Delta F \right| + \left| \frac{\partial (S)}{\partial d} \Delta d \right|$$

$$= \left| \frac{4}{\pi d^2} \Delta F \right| + \left| -\frac{8F}{\pi d^3} \Delta d \right|$$

$$\frac{\Delta S}{S} = \left| \frac{\Delta F}{F} \right| + \left| \frac{-2 \Delta d}{d} \right|$$

If the force applied is 100 lb_f and the diameter is $\frac{1}{4}$ in.

$$\frac{\Delta S}{S} = \left| \frac{1 \text{ lb}}{100 \text{ lb}_f} \right| + \left| \frac{-0.02 \text{ in.}}{0.25 \text{ in.}} \right| = 0.01 + 0.08 = 0.09$$

Also,

$$\frac{\Delta S_{\text{rss}}}{S} = \sqrt{(0.01)^2 + (0.08)^2} = 0.0806$$

Therefore, for this load and diameter the stress is

$$S = \frac{6400 \text{ lb}_f}{\pi \text{ in.}^2} \pm \Delta S_{\text{rss}} = 2038.2 \pm 164.3 \frac{\text{lb}_f}{\text{in.}^2}$$

or

$$S = 2038.2 \pm \Delta S = 2038.2 \pm (0.09)(2038.2) = 2038.2 \pm 183.4 \text{ lb}_f/\text{in.}^2$$

Since 2σ limits were used, 95 per cent of the data will be given by these limits.

For the above example, the major contributor to the error is the measurement of length (it contributes 8 per cent error). A more accurate solution of the problem can be obtained by measuring the length to ± 0.001 in. The decision to be made at this point in an experimental program is governed by whether or not the results are accurate enough to represent the problem. If the results are not satisfactory, the decision might be to eliminate some of the uncertainty by a more accurate measurement of the parameters that make the largest contribution to the error.

5.5
Analysis of
the Dynamic
Systems

Before any data-analysis discussion could be complete, it must contain methods for dealing with dynamic systems. For these systems the transient characteristics of the measurement system must be described. If this is the reader's first introduction to the field of measurements, it may be more practical to skip the discussion of dynamic systems until more experience has been gained. However, the section is essential to an overall understanding of measurement systems. In fact, an understanding of dynamic measurements gives the most general coverage of all measurement systems.

The dynamic characteristics of a measurement system can be represented in general operator terms as

$$(a_n D^n + a_{n-1} D^{n-1} + \cdots + a_1 D + a_0)I_o$$

where the a's are coefficients, the D's are derivatives, and I_o is the information output of the measurement system. The input information can also be represented in operator form:

$$(B_m D^m + B_{m-1} D^{m-1} + \cdots + B_1 D + B_0)I_i$$

where the B's are coefficients and I_i is the input information.

For the static or steady-state measurement, the operator form is

$$A_0 I_o = B_0 I_i$$

The ratio of output information over to information is called the gain of the system,

$$\frac{I_o}{I_i} = \frac{B_0}{A_0} = K$$

where K is the static gain. This term can be found from the static calibration of a measurement system. We have already covered the mathematical tools for finding this value and for making a statistical analysis of it. This type of system is a zeroth-order system. Two additional measurement systems are commonly used: first-order systems and second-order systems.

Without considering the input information at this time, the operator form for a second-order measurement system is given:

$$(A_2 D^2 + A_1 D + A_0)I_o = B_0 I_i$$

Dividing both sides by the constant, A_0, gives

$$(\bar{A}D^2 + \tau D + 1)I_o = KI_i$$

The purpose of this section is to develop methods for a statistical analysis of the new terms, τ and $\bar{A}$. The coefficient τ is called the time constant. Appendix III gives calibration methods for finding K, τ, and $\bar{A}$.

In static measurement problems, it is recommended that this section be skipped and that coverage be delayed until dynamic systems are considered. Dynamic responses are the foundations of control systems. One very simple electrical circuit is a convenient vehicle for introducing the fundamentals of transient response. This circuit is the inductance–capacitance–resistance circuit (also called the filtering circuit).

Filtering circuits are for use with a varying input signal. When a signal is produced with a frequency that is to be detected, it is possible that extraneous signals are also introduced by the sensor. Extraneous signals should not be reflected by the output signal. To accomplish this goal it is necessary to remove the extraneous inputs before recording. This task can often be accomplished by filtering circuits. When the input frequency is of a known value, the filtering circuits can be designed to avoid any problem of extraneous signals which have frequency variation from the desired input signal. This type of signal modification is very similar to the filtering circuit available in a radio or television tuning circuit. All extraneous signals are rejected by these circuits. The operation of a tuning circuit in either a radio or a television set is identical to the filtering methods employed in the measurement of input signals. The various types of filtering circuits are of considerable importance to the measurement system. The following analysis is the basis of the methods employed by the measurement engineer.

The mathematical representation of an electrical circuit can be considered in two forms. The procedure in this text will include equations in the forms of voltage, current, and charge. Consider the loop equation of the circuit of Figure 5.13.

$$e_i = R\frac{dq}{dt} + \frac{1}{C}q \tag{5.28}$$

$$e_o = \frac{1}{C}q \tag{5.29}$$

In filtering out undesirable frequencies, it is of interest to determine the ratio of the output signal to the input signal with variation in the frequency of the input signal. This may be accomplished by assuming that the input

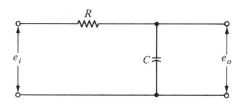

Figure 5.13 Low Pass Filter

signal is of the form $e_i = A \sin \omega t$. Now Equation 5.28 can be solved for a variable frequency input, ω, with boundary condition $t = 0$, $q = 0$ to give, in operator form,

$$(RCD + 1)q = CA \sin \omega t \qquad (5.30)$$

(where q is the output information), with the complementary solution

$$q = C_1 e^{-t/RC} \qquad (5.31)$$

for the given boundary condition $C_1 = 0$. Then, the particular solution may be obtained according to the following. Assume that

$$q = B \cos \omega t + E \sin \omega t$$

Then Equation 5.30 becomes

$$RC\omega(-B \sin \omega t + E \cos \omega t) + B \cos \omega t + E \sin \omega t = CA \sin \omega t \qquad (5.32)$$

Solving for B and E gives

$$B = \frac{-\omega RC^2 A}{\omega^2 R^2 C^2 + 1} \quad \text{and} \quad E = \frac{CA}{(\omega RC)^2 + 1} \qquad (5.33)$$

and

$$q = \frac{-\omega RC^2 A}{(\omega RC)^2 + 1} \cos \omega t + \frac{CA}{(\omega RC)^2 + 1} \sin \omega t \qquad (5.34)$$

Therefore,

$$e_o = \frac{1}{C} q = \frac{-\omega RCA}{(\omega RC)^2 + 1} \cos \omega t + \frac{A}{(\omega RC)^2 + 1} \sin \omega t \qquad (5.35)$$

Since a form $A \cos \omega t + B \sin \omega t$ can always be expressed in the form $C \sin(\omega t \pm \phi)$,

$$e_o = \frac{1}{\sqrt{(\omega RC)^2 + 1}} A \sin(\omega t - \phi) \qquad (5.36)$$

where

$$\phi = \tan^{-1} \omega RC \tag{5.37}$$

Now, the amplitude of the output divided by input voltage is

$$\frac{e_o}{e_i} = \frac{1}{\sqrt{(\omega RC)^2 + 1}} \tag{5.38}$$

and the output lags the input by the phase angle, ϕ. In operator form this is expressed

$$\frac{e_o}{e_i}(D) = \frac{1}{(RCD + 1)} \tag{5.39}$$

It is necessary to consider the effect of frequency on amplitude ratio and phase angle. To accomplish this, it is convenient to introduce the definition of decibel at this point. The decibel is a logarithmic power-amplitude ratio term,

$$\text{decibel} = \text{dB} = 10 \log_{10} \frac{P_o}{P_i} \tag{5.40}$$

where

$$P_o = \text{power out}$$
$$P_i = \text{power in}$$

In electrical units $P = EI$ and, if the input and output impedances are equal,

$$\text{decibel} = \text{dB} = 10 \log_{10} \frac{I_o^2 z_o}{I_i^2 z_i} = 20 \log_{10} \frac{I_o}{I_i} \tag{5.41}$$

$$= 10 \log_{10} \frac{E_o^2/z_o}{E_i^2/z_i} = 20 \log_{10} \frac{E_o}{E_i} \tag{5.42}$$

where both voltage, E, and current, I, are amplitude ratios only. The phase relationships are not considered at this point. Using Equation 5.42 and the results of the amplitude ratio, Equation 5.38, it is possible to determine the variation of amplitude with respect to frequency:

$$\text{dB} = 20 \log_{10} \frac{1}{\sqrt{(\omega RC)^2 + 1}} = -20 \log_{10} \sqrt{(\omega RC)^2 + 1}$$
$$= -10 \log_{10} [(\omega RC)^2 + 1]$$

When $\omega RC \ll 1$, the gain in decibels is:

$$\text{dB} = -10 \log_{10} 1 = 0$$

and when $\omega RC \gg 1$, the gain in decibels is:

$$dB = -10 \log_{10} (\omega RC)^2 = -20 \log_{10} \omega RC$$

A logarithmic plot is a graphical representation of amplitude (in decibels) versus frequency. When this is plotted for the simple RC circuit, the variation is shown in Figure 5.14. When the input frequency is greater than 100

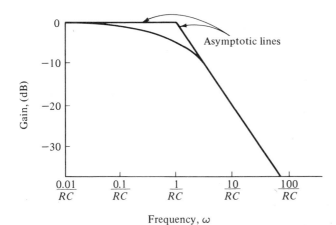

Figure 5.14 Typical Gain-Frenquency Plot

times the product RC, the amplitude of the output is one one-hundredth $(\frac{1}{100})$ the input. This means that the output is relatively unaffected by extraneous inputs having a frequency much higher than the product RC, compared to input frequencies that are at frequencies much lower than this product.

For example, suppose that $R = 0.1$ MΩ and $C = 0.10$ μF; then RC $= 0.010$. If the input signal is composed of two frequencies, of 5 and 10,000 Hz the amplitude of the 5-Hz signal would be

$$dB = -20 \log_{10} \sqrt{(0.05)^2 + 1.0} = 0$$

or

$$0 = 20 \log_{10} \frac{e_o}{e_i} \quad \text{and} \quad e_o = e_i$$

But the 10,000 Hz signal would have an amplitude ratio

$$dB = -20 \log_{10} \sqrt{100 + 1} = -20$$

or

$$-20 = 20 \log_{10} \frac{e_o}{e_i} \quad \text{and} \quad e_o = 0.1 e_i$$

Ninety per cent of the higher-frequency signal is filtered out of the output.

Two advantages of the operational form of differential equations applicable to measure systems are as follows:

1. Operators acting in series on an input signal can be simply algebraically multiplied.
2. Representation on a logarithmic plot is simple for sine–cosine types of input signals.

Before continuing the modifications available for filtering circuits, these two advantages will be exploited. In Figure 5.15, the operator notation is

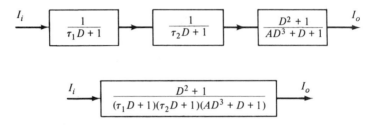

Figure 5.15 Equivalent Block Diagrams

used to obtain an equivalent block diagram. Without considering the proof of the above-stated equivalence, an explanation involving the techniques of solving differential equations will be used. The solution of the operator equation

$$(A_n D^n + A_{n-1} D^{n-1} + \cdots + A_2 D^2 + A_1 D + A_0) I_o$$
$$= (B_m D^m + B_{m-1} D^{m-1} + \cdots + B_2 D^2 + B_1 D + B_1) I^i \quad (5.43)$$

involves two parts, the complementary and the particular solutions. The complementary solutions to Equation 5.43 result from finding the roots of the corresponding algebraic equations when the operator, D, is replaced with the variable, r. Therefore, algebraic multiplication of the operators will not change the roots of the equation. Since these roots also govern the form of the particular solution in algebraic form, the particular solution is also unaffected by the simple multiplication of operation elements in series.

The use of operational form in determining the logarithmic plot of a measurement system will be demonstrated by using an input signal that is, again, trigonometric in form. This technique is a trigonometric transfer function for the operator. The left-hand side of Equation 5.43 represents the dynamic characteristics of the measurement system, while the right-hand side represents the fluctuations of the input signal. To obtain the complementary solution to this equation, the right-hand side is set equal to zero and the resulting differential equation is solved. One method of solution is to find the roots of the associated algebraic equation resulting when r is substituted for the operator, D.

$$A_n r^n + A_{n-1} r^{n-1} + \cdots + A_1 r + A_0 = 0 \qquad (5.44)$$

When $n > 3$, no method exists for absolute solution of the algebraic equation. (When $n = 1$, $n = 2$, or $n = 3$, the solution results from simple algebra, the quadratic equation, or the cubic equation.) When $n > 3$, two methods are employed: factoring and finding roots by the remainder theorem. At any point where the remaining operator form reduces to factors times a cubic or quadratic form, the system can be solved by applying the proper formula. For example,

$$r^4 + 2r^3 + 5r^2 + 8r + 4 = (r^2 - 4)(r^2 + 2r + 1)$$
$$= (r - 2)(r + 2)(r + 1)^2$$

and the roots have been determined. The complementary solution to the equation

$$(D^4 + 2D^3 + 5D^2 + 8D + 4)I_o = 0$$

is

$$I_0 = C_1 e^{-2t} + C_2 e^{2t} + C_3 e^{-t} + C_4 t e^{-t}$$

For each root of the algebraic equation there exists a solution that will satisfy the differential equation. The forms of the solution depend on the nature of the roots. A simple classification follows.

1. Roots are real and distinct: $r_1, r_2, r_3, \ldots, r_k$. The solutions are $C_1 e^{r_1 t} + C_2 e^{r_2 t} + \cdots + C_k e^{r_k t}$.
2. Roots are real and repeated: $(r - 1)^m$, $r_1 = 1$, $r_2 = 1, \ldots, r_m = 1$. The solutions are $C_1 e^t + C_2 t e^t + C_3 t^2 e^t + \cdots + C_m t^{m-1} e^t$.
3. Roots are complex and distinct: $r_1 = a_1 + ib_1$, $r_2 = a_2 + ib_2, \ldots,$ $r_j = a_j + ib_j$. The solutions are $C_1 e^{a_1 t}[\cos(b_1 t + \phi_1)] + C_2 e^{a_2 t}[\cos(b_2 t + \phi_2)] + \cdots + C_j e^{a_j t}[\cos(b_j t + \phi_j)]$.
4. Roots are complex and repeated: $[r^2 - 2ar + (b^2 + a^2)]^p$, $r_1 = a \pm ib$, $\ldots, r_p = a \pm ib$. The solutions are $c_1 e^{at}[\cos(bt + \phi)] + C_2 t e^{at}[\cos(bt + \phi_2)] + \cdots + C_p t^{p-1} e^{at}[\cos(bt + \phi p)]$.

Fortunately, two factors reduce the complexity of the complementary solution. First, the order of the differential equation of the measurement system is usually less than 4 (i.e., $n = 1, 2,$ or 3). Additionally, the boundary conditions for good measurement systems are controlled in such a way that solutions involving time to a power and positive exponentials of time have their coefficients equal to zero. The negative exponentials damp out with time. Another way of stating this fact is that the transient terms (i.e., solutions of the complementary equation) vanish when time increases. Therefore, in many cases the complementary solution may be neglected in terms of output information after a short transition period. Then the particular solution is essentially the total solution.

Now consider the particular solution and methods of determining its form. The method to be discussed here is the method of undetermined coefficients. This involves guessing a solution for I_o in the form of the total input signal and forcing the solution to solve the differential equation by selecting the correct value for the undetermined coefficients. When the input signal, I_i, is trigonometric in form, the solution is guessed using the same trigonometric functions with arbitrary coefficients. For example, when

$$I_i = A \cos \omega t$$

guess that $I_o = C_1 \cos \omega t + C_2 \sin \omega t$. A more general form of this method is to represent the signals in exponential form and allow the possibility of constants which are complex numbers (real and imaginary). Then

$$I_i = A e^{i\omega t} \tag{5.45}$$

and

$$I_o = C_1 e^{i(\omega t + \phi)} \tag{5.46}$$

To present a complete analysis of the transfer function, it is important to make two observations at this point. First, Equation 5.43 is classified as a linear differential equation. This means that any solutions found that satisfy a portion of the differential equation can be added together to form a composite or total solution. Therefore, if the input signal is composed of several different frequency trigonometric signals, a solution for each frequency can be found, and all of the solutions can be added to obtain a total solution. This is stated without proof, but it can be verified by studying any mathematics textbook on linear differential equations. Second, any periodic input signal that is continuous can be represented exactly by an infinite sum of sine and cosine functions with integral frequencies. This fact can be established by studying a textbook on Fourier series. Therefore, the analysis to be presented here can be applied to very complex input signals by duplicating the procedure to be outlined for each frequency involved in the input signal.

The procedure to be followed to solve for the input signal of Equation 5.45 is first to guess the form of the solution to Equation 5.46.

Now, performing the operations on these signals, I_i and I_o, indicated by Equation 5.43 gives

$$C_1[A_n(i\omega)^n + A_{n-1}(i\omega)^{n-1} + \cdots + A(i\omega) + A_0]e^{i(\omega t + \phi)}$$
$$= A[B_m(i\omega)^m + B_{m-1}(i\omega)^{m-1} + \cdots + B_1(i\omega) + B_0]e^{i\omega t} \qquad (5.47)$$

This is demonstrated by repeatedly taking the indicated derivatives,

$$DI_o = C_1(i\omega)e^{i(\omega t + \phi)}$$

where C_1, ω, and ϕ are constants.

$$D^2I_o = C_1(i\omega)^2 e^{i(\omega t + \phi)}$$

and

$$D^nI_o = C(i\omega)^n e^{i(\omega t + \phi)}$$

Substitution of these derivatives and similar ones taken on I_i into Equation 5.43 gives Equation 5.47. Now, Equation 5.47 is an algebraic equation and can be rearranged in the form

$$\frac{I_o}{I_i} = \frac{Ce^{i(\omega t + \phi)}}{Ae^{i\omega t}} = \frac{[B_m(i\omega)^m + B_{m-1}(i\omega)^{m-1} + \cdots + B_1(i\omega) + B_0]}{[A_n(i\omega)^n + A_{n-1}(i\omega)^{n-1} + \cdots + A_1(i\omega) + A_0]}$$
$$(5.48)$$

The complex number on the right-hand side of Equation 5.48 is the sinusoidal transfer function for the differential equation 5.43. For comparison, Equation 5.43 is written in operator form:

$$\frac{I_o}{I_i}(D) = \frac{(B_mD^m + B_{m-1}D^{m-1} + \cdots + B_1D + B_0)}{(A_nD^n + A_{n-1}D^{n-1} + \cdots + A_1D + A_0)} \qquad (5.49)$$

It should be noted that the sinusoidal transfer function can be obtained from the operator form of a differential equation by simply substituting $i\omega$ for the operator D.

The analysis of the complex number on the right-hand side of Equation

5.48 can be discovered by considering the term to which it is equal:

$$\frac{C_1 e^{i(\omega t + \phi)}}{A e^{i\omega t}} = \frac{C_1}{A} e^{i\omega t + i\phi - i\omega t} = \frac{C_1}{A} e^{i\phi}$$

$$= \frac{C_1}{A}(\cos\phi + i\sin\phi) \tag{5.50}$$

Expressed in polar form, Equation 5.50 becomes

$$\frac{C_1 e^{i(\omega t + \phi)}}{A e^{i\omega t}} = \frac{C_1}{A} \tag{5.51}$$

In terms of the measurement system, the sinusoidal transfer function (the complex number on the right-hand side of Equation 5.48) is equal to the gain of the measurement system, and the output signal lags the input signal by the angle ϕ. The following example demonstrates the usefulness of the above analysis.

EXAMPLE 5.5

Sinusoidal Transfer Function. Consider the filtering circuit of Figure 5.13, where $e_i = A\sin\omega t$. Then, using the operator form of this circuit,

$$\frac{e_o}{e_i}(D) = \frac{1}{RCD + 1}$$

The sinusoidal transfer function is

$$\frac{e_o}{e_i} = \frac{1}{RC(i\omega) + 1}$$

To obtain a real number in the denominator, multiply numerator and denominator by the complex conjugate of the denominator to obtain

$$\frac{e_o}{e_i} \frac{1(1 - RCi\omega)}{(RCi\omega + 1)(1 - RCi\omega)} = \frac{1 - RC\omega i}{1 + (RC\omega)^2}$$

Expressing this in polar form gives

$$\frac{e_o}{e_i} = \frac{1}{1 + (RC\omega)^2}$$

$$\phi = \tan^{-1}(\omega RC)$$

These results are the same as those obtained earlier by solving the differential equation. If this were the only application of the sinusoidal transfer function, the efforts would have been wasted. However, the utility of this technique can be realized in two very important ways: (1) the effect of one operator can be superimposed on the effect of a second operator in series, and (2) the phase angles resulting from two operators can be added algebraically. The following examples best display the advantages gained by employing transfer functions.

EXAMPLE 5.6

Consider the filtering circuit shown in Figure 5.16. The operator form of the first dotted block is the same as before:

$$\frac{e_{o1}}{e_i}(D) = \frac{1}{RCD + 1}$$

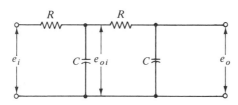

Figure 5.16 Sharp Cut-Off Filtering Circuits

Also,

$$\frac{e_o}{e_{o1}}(D) = \frac{1}{RCD + 1}$$

But operator algebra from Figure 5.15 allows

$$\frac{e_o}{e_i}(D) = \frac{1}{(RCD + 1)(RCD + 1)}$$

for the complete filtering circuit. A plot of the complete output can be made by adding the two responses graphically (see Figure 5.17).

To assist in plotting response curves some comments on the previous example are in order. When the operator is a first-order operator in the denominator, the true gain falls between two asymptotic lines. When

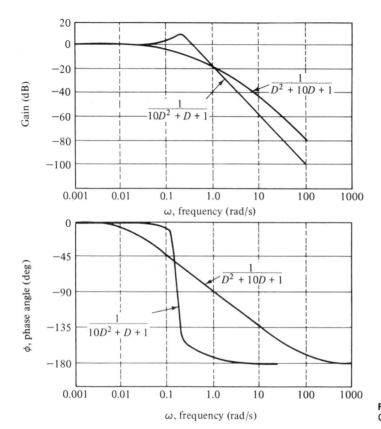

Figure 5.17 Second Order Response Curves

$\omega \ll 1/RC$, the gain approaches 0 dB. When $\omega \gg 1/RC$, the gain approaches the line with a slope of -20 dB/decade, where a decade is a change in frequency by a factor of 10. The point of intersection of these two asymptotic lines has the coordinates $(1/RC, 0 \text{ dB})$. The phase angle lags, with near zero lag when $\omega \ll 1/RC$. When $\omega = 1/RC$, the phase lag is 45°, and when $\omega \gg 1/RC$, the phase angle approaches 90° lag. When the operator is a second-order operator in the denominator, the asymptotic lines are the 0-dB line and the line with a slope of -40 dB/decade. The phase-angle lag is doubled at each point. When the denominator has two different roots, the individual curves can be drawn for each operator, and a final output can be determined by addition. This is extremely convenient for more complex systems.

Two items to remember when applying the transfer function are that the intersection of the asymptotic lines for the gain plot occurs at the value of frequency which makes the term ωRC equal to one. Second, this point is the point where the lag is 45°.

Before continuing the discussion of other filtering circuits, the effect of operators in the numerator will be shown.

EXAMPLE 5.7

Consider the operator

$$\frac{e_o}{e_i}(D) = \frac{AD + 1}{(RCD + 1)^2}$$

Since the effect of the terms in the denominator are already known, attention will be devoted to the numerator. The polar form of the numerator is

$$\frac{e_o}{e_i}(i\omega) = \sqrt{(A\omega)^2 + 1}$$

$$\phi = \tan^{-1}(A\omega)$$

For the consideration of gain,

$$dB = 20 \log_{10}\sqrt{(A\omega)^2 + 1}$$

When $A \ll 1/\omega$, gain $= 0$ dB; when $A \gg 1/\omega$, gain $= 20 \log_{10} A\omega$.

The asymptotic lines are 0 dB and the line with slope $+20$ dB/decade (for first-order operators in the numerator). The phase angle ϕ now leads the input signal, and the break point (the intersection point of the asymptotic lines) occurs when $A = 1/\omega$ when the phase angle leads by 45°. Now, if $A = RC$, the gain plot will be exactly like that of Figure 5.14, but the gain will be positive. If the operator in the numerator is also second order and identical to the denominator, the system will have a gain of 0 dB and zero phase angle for all frequencies. This then is a zeroth-order system. There is available a technique for aiding in the development of zeroth-order systems.

The basic technique for any operator system is simply to normalize the complex number which results from the substitution of $i\omega$ for D in the operator equation. However, if the operators are factored in the numerator or denominator, the contribution of each part can be determined from the foregoing analysis.

$$\frac{1}{RCD+1} \longrightarrow dB = -20 \log_{10} \sqrt{(\omega RC)^2 + 1}$$

$$\frac{RCD+1}{1} \longrightarrow dB = +20 \log_{10}(\omega RC)^2 + 1$$

$$\frac{1}{AD} \longrightarrow dB = -20 \log_{10} A\omega$$

$$\frac{AD}{1} \longrightarrow dB = 20 \log_{10} A\omega$$

Since

$$\log_{10} \frac{AB}{C} = \log_{10} A + \log_{10} B - \log_{10} C$$

then

$$\frac{BD}{(RCD+1)(AD)} \longrightarrow dB = 20 \log_{10} B\omega - 20 \log_{10}\sqrt{(RC\omega)^2 + 1}$$
$$- 20 \log_{10} A\omega$$

The addition of phase angles in polar form follows the same rules for addition.

By using the concepts of sinusoidal transfer functions, linear equations, and Fourier series, it is possible to explain very simply many of the elements of a measurement system in a concise way. Filter circuits can be considered by developing the operational form of the differential equation. This will be discussed next. However, it should be emphasized that each element of the measurement system can be readily adapted to a complete dynamic analysis by establishing the transfer function for this element. This is an extremely powerful tool that is very well suited for application to measurement systems. This technique can be exploited at every opportunity where dynamic systems are used. Therefore, it is mandatory that a complete understanding of the methods be accomplished now. Perhaps the importance of filtering circuits alone does not justify the extensive coverage given the subject in this book. However, the secondary purpose of this presentation is the demonstration of the tecnhiques of transfer function. Therefore, the analysis of several different types of filtering circuits will be undertaken.

Consider Figure 5.18(a). The only difference from the previous circuit is the element across which the output signal is measured. Solving for the

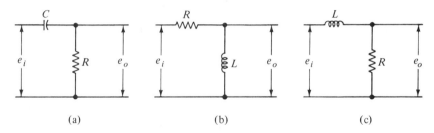

Figure 5.18 Various Filtering Circuits

operator form involves, first, writing the loop equations:

$$e_i = \frac{1}{C}q + R\frac{dq}{dt}$$

This equation has already been solved for $e_i = A \sin \omega t$, but now the output signal is

$$e_o = R\frac{dq}{dt} = R\frac{ARC^2\omega^2}{(\omega RC)^2 + 1}\sin \omega t + \frac{C\omega A}{(\omega RC)^2 + 1}\cos \omega t$$

$$= \frac{ARC\omega}{(RC\omega)^2 + 1}[RC\omega \sin \omega t + \cos \omega t]$$

$$\frac{e_o}{e_i} = \frac{RC\omega}{(RC\omega)^2 + 1}\sqrt{(RC\omega)^2 + 1}\frac{\sin(\omega t + \phi)}{\sin \omega t}$$

where

$$\phi = \tan^{-1}\frac{1}{RC\omega}$$

In operator form, the circuit of Figure 5.18(a) is given by

$$\frac{e_o}{e_i}(D) = \frac{RCD}{RCD + 1}$$

Solving the complex number to obtain the transfer function gives

$$\frac{RC(i\omega)}{RC(i\omega) + 1} = \frac{RCi\omega + (RC\omega)^2}{1 + (RC\omega)^2} = \frac{RC\omega}{\sqrt{(RC\omega)^2 + 1}} \qquad (5.52)$$

where

$$\phi = \tan^{-1}\frac{1}{RC\omega}$$

Figure 5.18(b) can be solved in similar manner by first writing the loop equations:

$$e_i = L\frac{di}{dt} + R_i$$

$$e_o = L\frac{di}{dt}$$

Without going into detail, the solution for the current can be found, when the input is $A \sin \omega t$, to be

$$i = \frac{-RL\omega}{R^2 + L^2\omega^2} \cos \omega t + \frac{R^2}{R^2 + L^2\omega^2} \sin \omega t$$

and

$$\frac{e_o}{e_i} = \frac{L\omega}{\sqrt{R^2 + L^2\omega^2}} \frac{\sin(\omega t + \phi)}{\sin(\omega t)} \qquad (5.53)$$

where

$$\phi = \tan^{-1} \frac{R}{L\omega}$$

For this problem the time constant $\tau = L/R$. For previous problems this time constant was RC. To recognize the similarities, the definition of time constant is introduced in the above analysis to obtain

$$\frac{e_o}{e_i} \frac{1}{\sqrt{1 + 1/\omega\tau}} \frac{\sin(\omega t + \phi)}{\sin \omega t}$$

where

$$\phi = \tan^{-1} \frac{1}{\tau\omega}$$

Therefore, Equations 5.52 and 5.53 reduce to the same result when the time constant has been properly defined. Both circuits perform the same function and have the same operational form. These circuits are called high-pass filters and have the operational equation

$$\frac{e_o}{e_i}(D) = \frac{\tau D}{\tau D + 1} \qquad (5.54)$$

The student should verify that Figure 5.18(c) is a low-pass filter with an operational equation identical to Equation 5.39 when the time constants have been properly defined.

The similarities in all of the circuits considered here (involving two elements) result from the fact that they all result in first-order systems. The proper definition of terms reduces the operator forms to two types. Corresponding similarities exist between the above electrical systems and the mechanical systems of the following example. The similarities are pointed out in Example 5.8 by a simple analysis of the governing differential equations.

EXAMPLE 5.8

Mechanical Systems. In Figure 5.19 four mechnical systems are shown along with the parameter τ and the operational equations. It is of importance to indicate the method of obtaining these operators. In part (a) the force

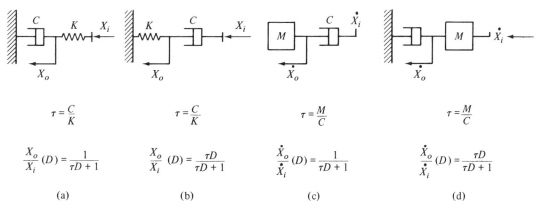

$$\tau = \frac{C}{K} \qquad\qquad \tau = \frac{C}{K} \qquad\qquad \tau = \frac{M}{C} \qquad\qquad \tau = \frac{M}{C}$$

$$\frac{X_o}{X_i}(D) = \frac{1}{\tau D + 1} \quad\quad \frac{X_o}{X_i}(D) = \frac{\tau D}{\tau D + 1} \quad\quad \frac{\dot{X}_o}{\dot{X}_i}(D) = \frac{1}{\tau D + 1} \quad\quad \frac{\dot{X}_o}{\dot{X}_i}(D) = \frac{\tau D}{\tau D + 1}$$

(a)   (b)   (c)   (d)

Figure 5.19 Mechanical Systems

transmitted through the spring equals the force dissipated in the damper. Then,

$$K(X_i - X_o) = C\frac{dX_o}{dt}$$

$$\frac{C}{K}\frac{dX_o}{dt} + X_o = X_i$$

In operator form,

$$(\tau D + 1)X_o = X_i$$

$$\frac{X_o}{X_i}(D) = \frac{1}{\tau D + 1}$$

where

$$\tau = \frac{C}{K}$$

In part (c) the force transmitted through the damper is the force causing the acceleration of the mass.

$$C\frac{d(X_i - X_o)}{dt} = M\frac{d^2X_o}{dt^2}$$

or

$$C\dot{X}_i = M\ddot{X}_o + C\dot{X}_o$$

Then

$$(\tau D + 1)\dot{X}_o = \dot{X}_i$$

and

$$\frac{\dot{X}_o}{\dot{X}_i}(D) = \frac{1}{\tau D + 1}$$

where

$$\tau = \frac{M}{C}$$

The next step in this analysis is to involve a second-order system that cannot be solved with real roots. For the system of Figure 5.20, the system is dependent on two parameters. The corresponding equations governing

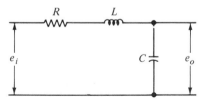

Figure 5.20 Three Element Filter

the circuit of Figure 5.20 are

$$e_i = L\frac{d^2q}{dt^2} + R\frac{dq}{dt} + \frac{1}{C}q \qquad (5.55)$$

and

$$e_o = \frac{1}{C}q \qquad (5.56)$$

When the complementary solution is reduced to zero, the particular solution for $e_i = A\cos\omega t$ is

$$q = \frac{A}{(R\omega)^2 + (1/C - L\omega^2)^2}\left[\left(\frac{1}{C} - L\omega^2\right)\cos\omega t + R\omega\sin\omega t\right]$$

$$e = \frac{1}{C}q = \frac{(1/C)A}{\sqrt{(R\omega)^2 + (1/C - L\omega^2)^2}}\cos(\omega t - \phi)$$

where

$$\phi = \tan^{-1} \frac{R\omega}{(1/C) - L\omega^2}$$

An expression for gain is

$$\frac{e_o}{e_i} = \frac{1}{\sqrt{(RC\omega)^2 + (1 - LC\omega^2)^2}} \angle -\phi \qquad (5.57)$$

$$\phi = \tan^{-1} \frac{CR\omega}{1 - LC\omega^2}$$

This form may be reduced to operator form by considering Equations 5.55 and 5.56 simultaneously:

$$e_i = LCD^2 e_o + RCDe_o + e_o$$
$$= (LCD^2 + RCD + 1)e_o$$

and

$$\frac{e_o}{e_i}(D) = \frac{K}{LCD^2 + RCD + 1} \qquad (5.58)$$

Verification that the sinusoidal transfer function applied to the above operator of Equation 5.58 gives the same result as that of Equation 5.57 should be made by the reader for practice.

Normally, a mathematical approach to higher-order systems is so complicated that an exact solution may not be obvious. Under these circumstances the mathematical solution may require so much time that it is impractical. However, an experimental solution may be obtained very readily by applying the electrical components to be used. This type of solution of differential equations is called the analog method of solution. The technique will not be pursued further in the present book. However, the technique does have considerable application in the analysis of measurement systems. This does not involve the solution of the basic differential equation, but it does include an understanding of the effective order of the measurement system. Here, effective order means the type of response a system may have (zeroth order, first order, etc.) in a particular application. The basic technique involves applying a known input signal, recording the output signal, and continuing for other known input signals. The characteristic curves for gain and phase angle are well known for first-order systems. After some experience with second-order systems, enough will be known to draw effective conclusions concerning the response of the measurement system to a given frequency input signal. This knowledge may extend to other types of inputs if the order of the measurement system can be determined.

At this time it is of interest to look again at second-order systems of the form of Equation 5.58,

$$\frac{I_o}{I_i}(D) = \frac{B_0}{A_2 D^2 + A_1 D + A_0} = \frac{K}{A_2^1 D^2 + A_1^1 D + 1}$$

$$dB = 20 \log_{10} \frac{I_o}{I_i} = 20 \log_{10} \frac{1}{\sqrt{(RC\omega)^2 + (1 - LC\omega^2)^2}}$$

$$= -10 \log_{10} [(RC\omega)^2 + (1 - LC\omega^2)^2]$$

Asymptotically, it is obvious that as $\omega \rightarrow \infty$

$$dB \longrightarrow -20 \log_{10} LC\omega^2$$

$$\longrightarrow -40 \log_{10} \omega(LC)^{1/2}$$

When $\omega \rightarrow 0$,

$$dB \longrightarrow -10 \log_{10} 1 = 0$$

For a particular example, let $RC = 1$ and $LC = 10$. Then Figure 5.21

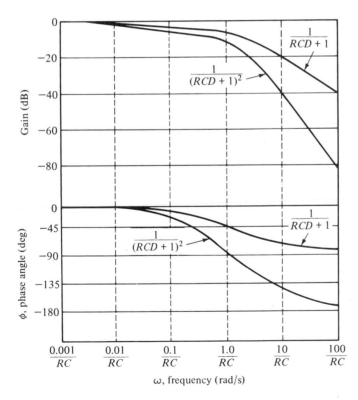

Figure 5.21 Response Plots of Sharp Cut-Off Filtering Circuit

gives the gain and phase-angle plots. Various other combinations are also shown in Figure 5.21. The point where the positive gain reaches its maximum in Figure 5.21 and where the phase angle becomes $-90°$ can be found mathematically.

To find the minimum value of

$$f(\omega) = [RC\omega)^2 + (1 - LC\omega^2)^2]$$

for variation in frequency, optimum values occur where the derivative of this quantity with respect to ω are zero.

$$\frac{df(\omega)}{d\omega} = 0 = 2(RC)^2\omega + 2(1 - LC\omega^2)(-2LC\omega)$$

Solving for ω gives

$$\omega = \sqrt{\frac{1}{LC} - \frac{1}{2}\frac{RC^2}{LC}} \qquad (5.59)$$

This is obviously a minimum value of the $f(\omega)$ given. For the particular example of Figure 5.21, this minimum occurs at $\omega = 0.308$. The value of ω for which the phase lag is $-90°$ occurs when the expression for phase angle, $\phi = \tan^{-1} RC\omega/(1 - LC\omega^2)$, has zero for its denominator. This occurs when

$$\omega = \sqrt{\frac{1}{LC}} \qquad (5.60)$$

For the particular example $\omega = 0.316$, Equations 5.59 and 5.60 can be very useful in the interpretation of experimental data collected from a second-order system.

Sufficient graphical and mathematical tools have been developed at this point to enable the reader to determine the coefficients of first- and second-order measurement systems [i.e., $(\bar{A}D^2 + \tau D + 1)I_o = KI_i$]. The value of K could be determined before reading Section 5.5. Now, τ and $\bar{A}$ can be determined experimentally for simple input signals. One method of evaluating statistically the parameters τ and $\bar{A}$ would be to determine their values experimentally 20 times or more. Then the techniques applied to static statistical analysis could be used.

When the input signal has a high level of noise superimposed on it (when signal/noise is less than 10), additional mathematical tools must be developed. The statistical methods of cross correlation and Gaussian distribution for an input signal having noise are presented in several references at the end of the chapter. Particular attention is called to reference 1.

5.6
Utility of the
Mathematical
Tools

Many applications of the mathematical tools developed in Sections 5.3 and 5.5 have already been made apparent by the previous discussion and examples. However, time must be taken to unify the concepts and to broaden the horizons of possible applications. The basic goal of any experimental program is set forth in the problem statement. Two examples will be discussed with respect to a reasonable application of statistical analysis.

EXAMPLE 5.9

The problem is to design a system to measure the thermal conductivity of a solid. The one-dimensional Fourier-type equation is found to hold for a circular shaft when no heat is transferred across the curved boundary, heat is added at one end of the shaft, and heat is removed at the other end. The equation is

$$\dot{q} = -KA\frac{\Delta T}{\Delta X}$$

where $\dot{q}$ is the heat-transfer rate, K is the thermal conductivity, A is the cross-sectional area of the shaft, and ΔT is the change in temperature over the length interval along the shaft, ΔX. To determine K experimentally, all of the boundary conditions required for the above equation must hold, and all parameters in that equation except K must be measured. The first condition can be satisfied by high-vacuum techniques, guard heaters, and radiation shields properly placed around the specimen. These precautions also facilitate the measurement of the heat-transfer rate, $\dot{q}$, which can be measured with a wattmeter if electrical heating is employed. Thermocouples are embedded to measure temperature. Area can be measured using a micrometer, and length increment can be measured using calipers and manometer.

How accurate can the measurement of thermal conductivity be? First,

$$K = -\dot{q}\frac{\Delta X}{A\,\Delta T}$$

Then

$$\frac{\Delta K}{K} = \left|\frac{\Delta \dot{q}}{\dot{q}}\right| + \left|\frac{\Delta(\Delta X)}{\Delta X}\right| + \left|\frac{-\Delta A}{A}\right| + \left|\frac{-\Delta(\Delta T)}{\Delta T}\right|$$

or

$$\frac{\Delta K_{rss}}{K} = \left\{ \left(\frac{\Delta \dot{q}}{\dot{q}}\right)^2 + \left[\frac{\Delta(\Delta X)}{\Delta X}\right]^2 + \left(\frac{\Delta A}{A}\right)^2 + \left[\frac{\Delta(\Delta T)}{\Delta T}\right]^2 \right\}^{1/2}$$

The numerator in each of the terms is the precision with which each parameter can be measured. The error in each case should be consistent, based on the same criteria:

1. Absolute error
2. Limited range data
3. Statistical error 1σ, 2σ, or 3σ limits.

If K must be bound so that no single measured value of K will be outside this bound, option 1 should be selected. Option 2 is usually present in all data, because designs are usually made for limited data ranges. Finally, if 80 per cent of the measured values of K are to be bound by the error limits, approximately 1.75σ limits should be selected. Let us that assume 95 per cent are to be bounded so that 2σ limits are selected.

Now, for this statistical limit, each error bound must be evaluated with the same limit. Two types of statistical data should be collected. (A sample with known thermal conductivity, length, area, and temperature is used to measure $\dot{q}$.) Plot true $\dot{q}$ versus measured $\dot{q}$ for the range of $\dot{q}$'s to be used in the system (have 20 points). Then, find the straight-line fit of the data using Equations 5.21 and 5.22. The values of $\dot{q}$ measured minus $\dot{q}$ from least squares (this is $\Delta \dot{q}$) must be checked by a chi-squared test for normalcy. If it is normally distributed, calculate the value of the standard deviation, σ_u, using Equation 5.8. Now, $\Delta \dot{q} = 2\sigma_u$.

The same type of technique would be used for determining the error in incremental temperature using known temperature inputs. However,

$$\Delta(\Delta T) = \Delta(T_2 - T_1) = |\Delta T_2| + |\Delta T_1|$$

Therefore, the error in measuring incremental temperature is twice as large as the error in measuring temperature.

Both length measurements are essentially single-input variables. (For this example we will neglect the effect of length change with temperature, linear coefficient of expansion.) Then 20 measurements of the distance between two thermocouples would be made. Mean value, standard deviation, and chi-squared calculations should be made on the data. The error in measuring area is developed from the area formula,

$$A = \frac{\pi d^2}{4}$$

and

$$\frac{\Delta A}{A} = \frac{2\,\Delta d}{d}$$

Again, 20 or more measurements of diameter are made using a micrometer. A chi-squared normalcy test is performed; after proving normalcy, the standard deviation is found using Equation 4.8. Then $\Delta d = 2\sigma_u$ and

$$\Delta A = \frac{2}{d_n}\left(\frac{\pi}{4}d_m^2\right)2\sigma_u = \pi d_m \sigma_u$$

EXAMPLE 5.10

An automatic fire extinguisher is to turn on a water spray system when any one of the thermocouple sensors are exposed to a flame. Evaluate the response time of a thermocouple sensor.

The best time to stop a fire is when it is initiated. The time required for the thermocouple to sense the presence of a flame and initiate the water spray is critical. It is also important that the spray is not initiated when no fire is present. Therefore, the dynamic output voltage of the thermocouple must be considered in the presence of a flame.

The experimental program is to start the sensor at room conditions and obtain the time required to provide an output voltage after it has been exposed to a flame. The sensor leads will be connected to an oscilloscope and suddenly (approximating a step change in temperature) a flame source (Bunsen burner) is placed under the sensor.

The heat transfer to the probe occurs by free convection,

(1) $$\dot{q} = hA(T_f - T_p)$$

where $\dot{q}$ is the heat-transfer rate, h is the heat-transfer coefficient, A is the probe surface area, T_f is the flame temperature, and T_p is the probe temperature. The heat added causes an increase in the probe temperature,

$$\dot{q} = m_p C \frac{d(T_p)}{dt}$$

where m_p is the mass of the probe, C is the specific heat of the probe material, and t is time. Then,

$$m_p C \frac{d(T_p)}{dt} = hA(T_f - T_p)$$

The governing equation is

$$\tau \frac{d(T_p)}{dt} + T_p = T_g$$

where

$$\tau = \frac{m_p C}{hA}$$

In operator form,

$$(\tau D + 1)T_p = T_g$$

and

$$\frac{I_o}{I_i} = \frac{T_p}{T_g} = \frac{1}{\tau D + 1}$$

The value of τ can be determined experimentally from a plot of output voltage versus time, since $T = f(V)$ is approximately a linear function for a limited temperature range. The solution of Equation (1) of this example is

$$T_p = C_1 e^{-t/\tau} + T_g$$

When $t = 0$, $T_p = T_{p_0}$, and $C_1 = T_{p_0} - T_g$.

$$T_p = T_{p_0} e^{-t/\tau} + T_g - T_g e^{-t/\tau}$$
$$= T_{p_0} e^{-t/\tau} + T_g(1 - e^{-t/\tau})$$

or with $\Delta T_p = T_p - T_{p_0}$,

$$\Delta T_p = T_g(1 - e^{-t/\tau}) - T_{p_0}(1 - e^{-t/\tau})$$
$$= (T_g - T_{p_0})(1 - e^{-t/\tau}) = \Delta T_T(1 - e^{-t/\tau})$$

where ΔT_T is the total change in the probe temperature. A plot of ΔT_p versus time is given in Figure 5.22. When $t = \tau$, the solution gives $\Delta T_p = 0.632$ ΔT_T. The experimental curve can be used at this point to determine τ for the first-order system. If the same test is run 20 times, a chi-squared test for τ can be made. Then the standard deviation can be obtained for normally distributed data. This information provides the dynamic characteristics of the sensor system.

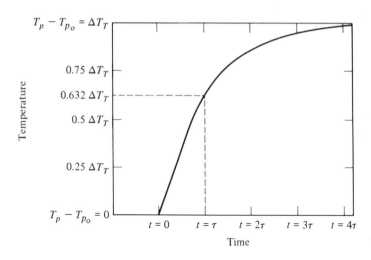

Figure 5.22 Response Curve of a Thermocouple

5.7
Conclusions

To obtain some limit on the uncertainty of precision, statistical methods are usually employed. For repeated measurements of the same parameter value with the same measurement system or with different measurement systems, a mean value and a standard deviation is calculated from Equation 5.8. Then the chi-squared test for normalcy is performed and statistical statements result; Student's "t" test can compare the probability that two measurement systems are equivalent in the case with different measurement systems.

The same type of evaluation is proposed when different individuals make the same measurement, and when the same measurement is made on seemingly identical models. Again, Student's "t" test can be applied to test the consistency of the sample space.

When the input parameter to be measured varies, the functional relationship between output and input ($I_o = f(I_i)$)can be fitted to a least-squares polynomial curve. When the curve is a stright line, Equations 5.21 and 5.22 are proposed for convenience. Then the deviation of the data point (ordered pair) from the least-squares curve is calculated using Equation 5.8; some test for normalcy must also be made.

When the measurement system must respond to a dynamic input, the basic coefficients of the system operator equation must be determined. Usually, this is an experimental process similar to that of Example 5.10 and Appendix III. A total of 20 evaluations of these coefficients can be used in a statistical analysis of their numerical value.

A statement of program results which contains no probability or absolute limits is inadequate. Data analysis may result in any one of four decisions:

1. The measurement systems must be designed to provide more precise values.
2. The experimental program hypothesis is apparently incorrect at some point (mathematical model, control parameters, physical model, etc.).
3. The program is adequate for the problem being considered.
4. The program provides a basis for expanding the basic hypothesis to other physical cases.

Experience and foresight can be very valuable in avoiding errors in the design of the experiment and in the design of the measurement system. Once the final application of the data is known, limits on precision can be made using Equation 5.26 or 5.27 before designing the measurement system. The capabilities of the measurement system might create a need to reconsider the control parameters.

The elements of experimental analysis (design of the experiment, design of the measurement system, and program analysis) are inherently related to each other. The selection of a plan in one area imposes a constraint in another area. The program goal can be attained more easily if the overall problem is reviewed before final decisions are made. For this reason it is important to have a clear understanding of the basic program elements and their interrelationship with each other. The conclusions of Chapters 2 and 3 should be reviewed at this time to reinforce the basic understanding of experimental analysis.

References

1. DOEBELIN, E. O., *Measurement Systems: Application and Design*, Mc-Graw-Hill Book Co., New York, 1966.

2. HALD, A., *Statistical Theory with Engineering Applications*, John Wiley & Sons, Inc., London, Chapman & Hall, Limited, 1952.

3. HOLMAN, J. P., *Experimental Methods for Engineers*, McGraw-Hill Book Co., Inc., New York, 1966.

4. *Precision Measurement and Calibration Statistical Concepts and Procedures*, NBS Serial Publication 300, vol. 1, U.S. Department of Commerce, Washington, D.C., 1969.

5. ABRAMOWITZ, M. and STEGUN, I. A. (eds.), *Handbook of Mathematical Functions*, NBS publication AMS 55, U.S. Department of Commerce, Washington, D.C., 1964.

6. SCHENCK, H., JR., *Theories of Engineering Experimentation*, 2nd ed. McGraw-Hill Book Co., Inc., New York, 1968.

PROBLEMS

5-1. Twenty individuals measure the length of a table using the same ruler and obtain 4ft 1in., 4ft $\frac{1}{2}$in., 4ft 2in., 4ft$\frac{1}{4}$ in., 4ft $\frac{1}{8}$in., 4ft 3in., 4ft 2in., 4ft 1in., 4ft $\frac{3}{4}$in., 4ft $\frac{3}{8}$in., 4ft $\frac{7}{8}$in., 4ft 1in., 4ft $\frac{7}{8}$in., 4ft $\frac{1}{2}$in., 4ft $\frac{3}{4}$in., 4ft $\frac{5}{8}$in., 4ft 1in., 4ft $\frac{1}{8}$in., and 4ft 2in. Plot the bar graph. Are the individual errors unbiased? What is the accuracy $\pm 3\sigma$ limits? Is the data normally distributed? What is the true length?

5-2. Two measurements of angular volocity are 101 rev/s and 101.1 rev/s. What statistical statements can be made concerning this data? Are the measurement systems equivalent? What is the true angular velocity?

5-3. Fit the data of Problem 5-1 to a normal distributation. What is σ?

5-4. The total thrown on 100 rolls of one die are

Total	Occurances
1	15
2	18
3	14
4	18
5	17
6	18

Is the die a true die? What is the σ limit on a single total? What is the probability that the number 1 will show on the next toss?

5-5. If 3600 throws of the dice give the total of Figure 5.3, what are the statistical statements that can be made concerning the experiment? Does the data satisfy the chi-squared test for normalcy?

5-6. In performing a calibration test of a copper–constantan thermocouple, the following readings were obtained for a 212°F water bath with an ice-point reference junction:

Number of Occurrences	Millivolts
1	4.511
1	4.483
3	4.439
4	4.400
4	4.379
10	4.300
4	4.221
3	4.200
2	4.161
2	4.117
1	4.089

Assume that the data is normally distributed. (a) Find σ. (b) If the true reading is 4.300 mV, find the accuracy in statistical terms. (c) Check for normalcy. (d) Plot the deviation data on probability paper to verify that the data is normally distributed.

5-7. In determining the melting point of a compound, eight measurements are made by two different analysts:

Analyst 1	Analyst 2
64.4°F	63.5°F
69.7°F	62.8°F
61.8°F	60.7°F
68.7°F	61.5°F

(a) Apply Student's "t" test to determine whether or not the individuals bias the data. (b) If each analyst is unbiased but each uses a different measurement system to obtain the above data, are the measurement systems unbiased? (Assume that each analyst's data is normally distributed).

5-8. Let each individual in class measure the diameter of a shaft along a given scribed mark using a ruler, a micrometer, and a caliber-steel tape system. Each individual is to measure the diameter five times using the sequence of measurement systems described. Are the students unbiased? Are the measurement systems equivalent?

5-9. Now use the micrometer to measure the diameter at 30°, 60°, 120°, and 150° rotation to the diameter measured in Problem 5-8. Collect sufficient data to evaluate the "roundness" of the shaft. Is the shaft statistically circular?

5-10. Ten measurements of angular rotational speed are to be made using four measurement systems: tachometer, strobotach, magnetic pick-up–counter, and magnetic pick-up variable-frequency–oscillator oscilloscope (Lissajous pattern). Are the measurement systems equivalent? If different individuals operate each system, are the individuals unbiased?

5-11. Find the least-squares fit to the data of millivolts versus temperature for copper–constantan thermocouples in the range 100–200°F for (a) a linear curve, (b) a parabolic curve. (Assume no variation in temperature, use thermocouple tables.)

5-12. Are the data from tables normally distributed around the linear curve determined in Problem 5-11? The parabolic curve?

5-13. Find an expression for the constants a, b, and c (of $y = ax^2 + bx + c$ with no variation in x) for the least-squares fit of four ordered pairs (x_1, y_1), (x_2, y_2), (x_3, y_3), and (x_4, y_4).

5-14. The specific volume of the gas in a tank is to be evaluated based on the measurement of pressure and temperature. If the pressure is measured with absolute accuracy of ± 4 per cent and temperature is measured with absolute accuracy of ± 5 per cent, what is the absolute specific volume of an ideal gas? If the pressure and temperature are accurate within 1σ limit accuracy $p \pm 1$ psia for 1σ limit and, $T \pm 2°F$ for 1σ limit, what is the statistical limit on specific volume.

5-15. The first law of thermodynamics is applied to a pump with insulated walls:

$$w = m(h_2 - h_1)$$

If $w = 10$ hp ± 0.1 hp absolute, $T = 60°F$ absolute, $p_1 = 14.7$ psia absolute, $p_2 = 100 \pm 1$ psia absolute, and $m = 10$ lb$_m$ absolute, what is the error in h_2?

5-16. If the work required to compress a spring is $w = F\,\Delta x$, what error in measuring F is acceptable (when the error in measuring Δx is ± 0.1 in. absolute) if the work is to be accurate to ± 0.01 ft-lb$_f$? (Assume that the measured values are $w = 25$ ft-lb$_f$, $f = 10$ lb$_f$, and $\Delta x = 2.5$ in.)

5-17. A thermocouple is a first-order dynamic system. Evaluate the time constant on a storage oscilloscope 10 times and determine the statistical time constant (a) in water, (b) in air.

5-18. A cantilever beam with strain gage is a second-order system. Evaluate the frequency of oscillation and damping constants 10 times and determine the statistical values of these constants in the operational equations for an initially displaced, zero-velocity beam.

5-19. Plot the gain–frequency plot for a sinusoidal temperature change with a thermocouple with time constant = 0.1 s.

5-20. Plot the frequency response of the cantilever beam of Problem 5-18 if the fixed end vibrates with various input sinusoidal displacements. (Assume a value of the constants in the operator equation if they have not been determined experimentally.)

5-21. Give the electrical analogy for a dynamic thermocouple.

5-22. Give an electrical analogy for the cantilever beam.

6

MEASUREMENT SYSTEMS FOR BASIC DIMENSIONS

6.1 Introduction

The definition of basic dimensions has historically been the starting point for the development of a technology in any scientific area. A basic definition of length was a prerequisite to the Egyptians' capabilities for building the pyramids. A certain amount of liberty is granted in defining dimensions, but constraints must be imposed if the defined parameters are to be consistent with the physical laws that have been developed. In the domain of measurements, the constraints are fully realized. Since all definitions of dimensions must also satisfy the physical laws, one is at liberty to define length; but to define another dimension for area one must be aware of the physical law constraining the relationship between length and area. Area is, in physical law, functionally dependent on the definition of length. (The area of a perfect square is defined to be the product of the length of the sides of the square.) Therefore, if the length of a platinum bar at a given temperature is defined to be 1 m, and the area of a perfect square with each side equal in length to this bar is defined to be 1 plot, the functional relationship between area and length requires 1 plot to be equal to 1 m times 1 m (or 1 m²). It is apparent that any other definition of area, with the previous definition of length, would be inconsistent with the physical law.

Based on the above analogy, it is possible to select dimensional quantities which are independent of each other. One such dimensional system results from the definition of the dimensions length, mass, and time. These three dimensions can be used with Newton's second law to establish the force dimension, $F = ML/t^2$. (Velocity, L/t, and acceleration, L/t^2, are derivable from two of these basic dimensions: length and time.) In fact, every thermodynamic dimension can be determined from these three basic definitions. (Pressure $= F/L^2$, specific volume $= L^3/M$, specific enthalpy $= FL/M$, etc.) Temperature is a manifestation of molecular activity, current is the numerical flow rate of electrons, and voltage is the presence or absence of electrons. The complex relationship between the basic defined dimensions (length, mass, and time) and temperature requires that some uncertainty be present in the determination of temperature from these dimensions (the Heisenberg uncertainty principle). The same complexity exists in determining dimensions for current and voltage. Therefore, these three dimensions are defined to be basic dimensions.

When a dependent dimension is defined without the use of the dimension on which it depends, the possibility of conflict between physical laws and dimensional systems exists. One way to avoid this conflict is to provide points of consistency. For the case of length and area, 1 plot was required to be 1 m². For temperature, current, and voltage, any inconsistency between physical laws and dimensional systems is alleviated by a similar definition to that used for area. The defining laws for these dimensions do not have the simplicity of length–area, but the net result is a dimensional system that is consistent with the physical laws.

For the present chapter, the basic dimensions to be considered are length, mass, time, temperature, current, and voltage. Measurement systems capable of sensing these dimensions will be discussed, and the basic standards used for their definition will be considered.

Before the disccussion of measurement systems and standards is begun, another possible method of defining the basic dimensions will be mentioned. This method involves defining length, mass, time, and force. Then, to be consistent with physical laws (i.e., $F \propto ML/t^2$), a constant of portionality must be found.

$$F = KM\frac{L}{t^2}$$

and

$$K = \frac{Ft^2}{ML}$$

This type of definition requires that many physical laws include a proportionality constant. Some dimensional systems use this technique. However, we will consider the previous postulation of defining dimensions. These dimensions will include only length, mass, time, temperature, current, and voltage. The accuracy of all measurements is ultimately determined by the accuracy with which these basic dimensions can be obtained. Therefore, this chapter will include the limits of these accuracies.

6.2
Measurement
Systems for
Length

The only requirements for defining a set of independent dimensions is that the dimensions be well defined. Recall that a modern-day standard has two requirements: the standard must be well defined, and the standard must be commonly accepted. However, a fundamental dimensional system requires only the first of these. One of the early standards for length was the distance from the elbow to the finger tip of the Egyptian pharaoh's right arm. This length was well defined; however, when a pharaoh died or when the pharaoh was a growing child, the standard changed. This chaotic state of standards would be totally unacceptable to the present-day individual. At a time when measurement within ± 1 in. was acceptable, the pharaoh's arm was adequate. After a day of stress, the arm dimension might be longer than at dawn. When measurement capabilities became more precise, the need for reproducible standards became much more desirable.

The National Bureau of Standards (NBS), under the Department of Commerce, has assumed the role of establishing standards for use in the United States. The United States first adopted an international standard for the measurement of length in 1893. The length was defined to be the distance between two finely scribed marks on a platinum–iridium bar maintained in a fixed environment and located at the International Bureau of Weights and Measures in Sèvres, France. This length was defined to be 1 m. Since that time the measurement capabilities have exceeded the definition of length because of the width of the scribed marks. In 1960 the defined length was further implemented by the definition of the meter in terms of the wavelength of the light radiated, through a vacuum, for transition of the krypton atom between the levels of $2p_{10}$ and $5d_5$. From this new definition,

$$1 \text{ m} = 1{,}650{,}763.73 \text{ wavelengths}$$

Originally, 1 m was defined to be equal to 39.37 in. by the NBS. It has been common usage to assume the appropriate relationship

$$1 \text{ in} = 2.54 \text{ cm}$$

This results in the value

$$1 \text{ in.} = 2.54 \text{ cm} = 0.0254 \text{ m} = (0.0254)(1{,}650{,}763.73) \text{ wavelengths}$$
$$\doteq 41{,}929.399 \text{ wavelengths}$$

Using the above definitions, it is possible to divide the inch into approximately 41,929 parts. With this standard, it is possible to measure length with an accuracy of $\pm$ 0.001 part in 41,929.399. Or length can be theoretically expressed:

$$1 \text{ in. } \pm 2.4 \times 10^{-8} \text{ in.}$$

If and when measurement systems can be devised which are capable of measuring more accurately than ± 0.024 micro-in., another standard for length must be defined. Until that time, the above definition will be sufficient.

Actual laboratory needs dictate the accuracy required for a specific measurement problem. With the present definition of length, any laboratory could have available (by purchasing a krypton lamp and associated equipment) a primary standard. This was not possible under the previous standard. Generally, the NBS provides a service of calibrating and certifying a set of secondary standards for almost any dimension required in the laboratory. Most engineering problems presently do not require the precision available. Therefore, certified secondary standards (secondary standards may be compared to the best standards available to the NBS) are frequently more than adequate for the normal problems encountered in engineering.

A general rule of thumb for any calibration process is that the standard (or secondary standard) have an accuracy that is 10 times that of the instrument being calibrated. While a wooden ruler with $\frac{1}{16}$-in. divisions might be sufficient for carpenters and some engineering lengths, more accurate measurement systems are required by machinists and the stress–strain analysts. The question of what precision is required is always answered by data analysis of a particular experimental program. This link binds together the components of experimental analysis: design of the experiment, design of the measurement system, and data analysis.

Measurement systems that are accurately calibrated are the nemesis of the fisherman. The use of the (frequently expanding) distance between two hands for determining length might adequately define the size of a fish, but a good carpenter requires a more precise method of defining length. The precision required increases with machining, strain measurements, and lattice structure problems. The length-measurement systems considered here are static measurement systems (i.e., the length will not change with time, as might not be the case with some fishermen). Although some measurement systems may be capable of measuring a dynamic length, the operation of these systems in the dynamic mode will be considered in the next chapter. The present section then will be concerned with static length measurement systems.

TAPE AND SCALE MEASURING SYSTEMS

The use of a solid material with or without scale divisions has been a common method of measuring length for ages. The simplicity of this measurement system should not be overlooked. However, the user is cautioned to remember that no measurement system should be used before some form of calibration has been performed. Although the major scale divisoins on a ruler or yardstick are usually in inches, the metric method of measure might occasionally occur. This means that some reliable source should be used to assure that the divisions are accurate enough for the purpose of their use. (Gage blocks, micrometers, steel machinist scales, or other sources should be used. Remember that the secondary standard used in calibration should be 10 times more accurate than the device being calibrated.)

The ruler or steel tape compares the unknown length to the divisions available on the tape. The care taken in this comparison will vary with the individual. The system is capable of measuring length if it is properly calibrated and used. However, extraneous inputs are present in the system in the form of temperature, parallex, and stress. (Other inputs in the form of pressure, humidity, refraction, etc., are neglected, since these effects are usually very small compared to the principal ones previously mentioned.) The list is sufficient to show that the output information is influenced by extraneous inputs.

Since most surveying is done using steel tapes, the utility of this type of measurement system cannot be ignored. The errors introduced by this measurement process might also include the location of the endpoints, sag in the tape, and stress applied. Temperature is always important when large lengths are measured.

When the same tape is used to measure a refrigerator and the space into which it is to fit, the tape becomes the standard for the situation and errors in absolute measurement are not important. Engineering problems must be reproducible in other localities using other measurement systems. Therefore, all extraneous inputs must be considered when data of this type are to be transmitted. Engineering scales are adequate for engineering drawings, because the same scale is used for all lengths (therefore, it is the standard) and the dimensions are placed on the drawings.

It is important to understand when a measurement is adequate and when it must be supplemented with correction factors. When the dimension is to be measured using the same system (adequately precise) and when the information is not to be distributed externally, the system can function properly. For other situations, the measurement system must be standardized (calibrated).

Most tape-measure or scale-type measurement systems are limited in accuracy to ± 0.01 in. (± 0.0254 cm). This is not an absolute limit, but the use of the human eye is practically limited to this range. When multiple applications of this type of system are made, the total error could be an accumulation of the individual errors in the most undesirable case.

GAGE BLOCKS

When additional accuracy is required, the principal secondary standard available in most engineering laboratories is the gage block. Gage blocks are rectangular masses of material with thicknesses that have been certified by the supplier or by the NBS. The accuracy of these individual blocks can be $\pm 2 \times 10^{-6}$ in. (5.08×10^{-6} m). The use of calibrated blocks (secondary standards) in the definition of a measurement system could provide an accuracy in the system of $\pm 20 \times 10^{-6}$ in. (50.8×10^{-6} cm). Gage blocks could be used for measuring a dimension by the methods of Figure 6.1.

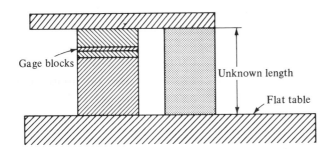

Gage blocks

Unknown length

Flat table

Figure 6.1 Gage Block Measuring System

The same measurement technique could be used to measure the diameter of a cylinder. However, the thickness of a gage block would typically be used to calibrate some additional measurement system, and the resultant system would be used to provide a workable secondary standards for the calibration of other systems. The optimal use of gage blocks rests on the ingenuity of the user. It should be realized that the resultant use of the blocks could actually have an error that is 10 times the error of the basic blocks (i.e., $\pm 20 \times 10^{-6}$ in.).

The principal use of gage blocks is to calibrate some length-measuring system. In ordinary laboratory work the accuracy would be sufficient. Temperature is still an extraneous input signal for the gage block. When the situation arises in which more accurate length measurements are required, a direct certification by the NBS is recommended.

MICROMETER, CALIPER, AND DIAL INDICATORS

The micrometer is an instrument known to any machinist. The physical principle of operation is the division of the screw threads on a shaft. The threads are ideally cut so that there are a fixed number of threads per inch of shaft (40 threads/in., 50 threads/in., or 100 threads/in.). If the threads are accurately cut, a tap will advance $\frac{1}{40}$ (or $\frac{1}{50}$, $\frac{1}{100}$) in./revolution. If one revolution is divided into 100 equal parts, it will be possible to divide 1 in. into 4000 parts (or 5000, 10,000). The accuracy of any micrometer depends on the accuracy with which the threads are cut and the accuracy with which the 360° circle is divided. If a 40 threads/in. shaft is turned, 1 revolution will represent 0.025 in. If the circle is divided into 25 equal parts, it is possible to measure length ± 0.001 in. (.00254 cm). When the shaft has 100 threads/in. and 100 divisions of the circle, the length can be measured ± 0.0001 in. (0.000254 cm). In either case the micrometer should be calibrated over its entire range using gage blocks. This will eliminate any error in machining and will give a 1-point temperature calibration.

One additional problem should be identified in association with micrometers. When a tap or nut is advancing in one direction on a screw thread, the advance is controlled by the machining process. However, when the tap advances and recedes, the clearance between screw thread and tap thread causes an error. This error can be essentially eliminated by measuring with a given thread–tap motion in each case. The machining clearance is an extraneous input for cases in which the tap advances and recedes in one measurement. Figure 6.2 shows a micrometer. The spindle moves with rotation of the thimble. The distance between the spindle and the anvil is indicated by marks on the barrel and the thimble. The thimble is threaded, and the barrel is tapped to measure the motion of the thimble relative to the barrel. The reading should be zero when the anvil is in contact with the spindle. When a micrometer is used to measure an unknown dimension, the stress applied to the unknown length can cause an extraneous input. The individual using the measurement system must consider this stress in the operation of the micrometer. The effect of nonparallel forces on the unknown length must be evaluated separately.

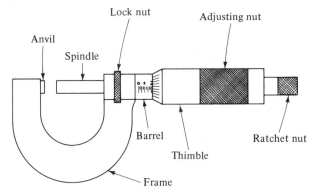

Figure 6.2 Micrometer

The calipers may measure either inside or outside dimensions, as shown in Figure 6.3. The length measured by the calipers is transferred to a micrometer or gage-block system for evaluation. The calipers are, therefore, a

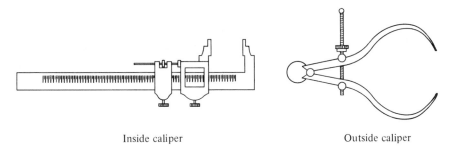

Inside caliper   Outside caliper

Figure 6.3 Calipers

method of transferring an unknown dimension to a measurement system capable of comparing the unknown dimension to a secondary standard. When calipers are used in a measurement process, the same extraneous inputs are present as are involved in using the micrometer (temperature, machining, clearance, and thread cutting).

A dial indicator operates on the same principle as the micrometer and calipers. The mechanical method involves the use of a rack and pinion. The linear displacement of the rack results in a rotational displacement of the pinion (see Figure 6.4). All of the extraneous inputs are the same as with the micrometer and calipers. Therefore, all three devices are sensitive to machining, clearance, and temperature effects. All these devices operate on the same physical principle and must be evaluated on the basis of this

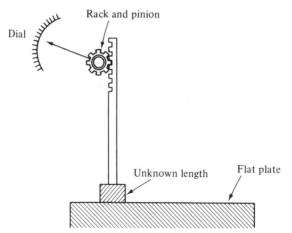

Figure 6.4 Dial Indicator

principle. These three devices could be accurate $\pm 1/10{,}000$ in. (0.000254 cm). Many engineering problems are adequately described using this accuracy.

MECHANICAL LINKAGE SYSTEMS

All of the mechanical measuring systems for length can employ additional linkage systems to amplify the length being measured. Two commonly used linkage systems are the levers of Figure 6.5 and the vernier scale of Figure

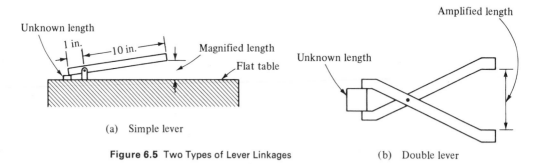

(a)   Simple lever

Figure 6.5 Two Types of Lever Linkages   (b)   Double lever

6.6. When the dimensions on the lever system are known (or certified by the proper agency), the relationship between the unknown length and the measured length is established. Usually, an amplification of 10 is the upper limit [i.e., measured length = (10)(unknown length)]. The vernier scale is used to divide a basic scale division into additional parts. In Figure 6.6

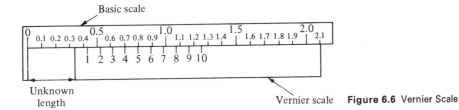

Figure 6.6 Vernier Scale

the vernier scale is for use with a scale having divisions of $\frac{1}{10}$ in. The total length of the vernier scale is $\frac{9}{10}$ in. and is divided into 10 equal parts of 0.09 in. each. If the vernier zero is at 0.35, the vernier 1 is at 0.44, the 2 is at 0.53, the 3 is at 0.62, the 4 is at 0.71, and the 5 is at 0.8. Here the vernier and basic scale match, and the scale plus vernier read 0.35. If the vernier were at 0.38, the match would occur with vernier 8 at 1.1. Therefore, when scale marks match with vernier marks, the fraction is determined on the vernier. Both devices, lever and vernier, essentially add one significant figure to the total measurement system. Gage blocks can easily be used to calibrate the linkage systems.

Many other varieties of linkage systems are available when more bars are employed. Angular-to-linear motions are available (in addition to the dial indicator). Stresses on the unknown length increase with amplification. The effect of temperature must be considered, and inertia becomes a problem with linkage systems when changing length is to be measured.

Since length is inherent to many physical laws, a very wide variety of length-sensing devices are possible. (The law of attraction, $r^2 = KM_1M_2/F$, could be used to determine length, r. However, this would require a constant, K, and measuring two masses and one force.) Here we will restrict the discussion to the most prominent length-measuring devices, with the understanding that the possibilities are almost unlimited.

OPTICAL METHODS FOR LENGTH MEASUREMENT

The use of optical techniques in measurements has attained a prominent position for several reasons. A single-wavelength source can be used to measure length with an accuracy of $\frac{1}{2}$ wavelength. Optical systems usually have a minimum loading effect in comparison to all other measurement systems. (They obey the first axiom of measurement, but they change the parameter being measured, usually by an insignificant amount.) Optical systems are essentially frictionless and inertialess. They are capable of responses up to the speed of light.

Because of the capabilities of this type of system, a major trend in refining measurements has been toward the use of optical systems. More details of the developments in this area are discussed in Chapter 9. The present chapter will consider only five basic systems: transits, microscopes, inferometers, monochrometers, and photographs.

TRANSITS

The surveyor's transit is a high-power telescope with crosshairs and gear trains for measuring angular displacements in the horizontal and vertical planes. A schematic diagram is shown in Figure 6.7. Leveling screws and

Figure 6.7 Schematic Diagram of a Transit

two mutually perpendicular level indicators are first used to establish the horizontal. Then a scaled target can be viewed at some unknown distance from the center of the transit. A pair of vertical angles and simple trigonometry can be used to obtain the distance from the transit to the target (see (Figure 6.8.) The accuracy can be affected by several factors. A plumb bob is usually used to establish the center of the transit. Therefore, air currents, vibrations of the earth, and methods of attaching the plumb bob could cause errors. The target whould have level indicators to establish the vertical position. The scale on the target could be in error. Machining errors are present in the transit. Finally, if the target is below the transit level (say in a valley) the presence of fog or thermal stratification of the air could change the index of refraction of light waves passing through the medium.

Two methods of checking accuracy are to shoot a series of at least three

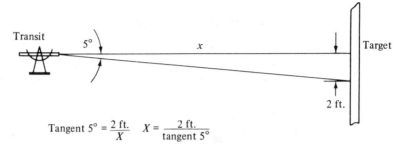

$$\text{Tangent } 5° = \frac{2 \text{ ft.}}{X} \quad X = \frac{2 \text{ ft.}}{\text{tangent } 5°}$$

Figure 6.8 Distance Determination Using a Transit

readings, with the last shot taken back at the initial starting position of the transit, and to shoot several known (usually taped) lengths. Additional equipment can be added to the transit to provide accuracies of ± 0.001 in. (0.00254 cm). These will be discussed in Chapter 9.

MICROSCOPES

Microscopes used for measuring length usually magnify the work piece in the low-power (50–100x) range. The actual measurement takes place in different ways, depending on the type of microscope empoyed. In one type the scale is permanently etched on the reticle of the microscope, and calibration is accomplished by viewing a certified scale. This is called a fixed-scale microscope.

A second type of microscope has a single etched line and is mounted on a micrometer-type (screw thread and tap) slide. The measurement is only as accurate as the machining and marking of the micrometer. Calibration could again be performed by viewing a certified scale.

The filar microscope (or bifilar microscope) has one (or two) etched marks that can be positioned using an external micrometer. The distance moved from end of the measured object to the other end is read on the micrometer dial. Accuracy is in the range of ± 0.001 inch (0.00254 cm).

The limits on accuracy for all of these measuring microscopes are specified by the manufacturer. The greatest errors in using these systems occur from improper focusing. Experience with a calibrated standard is the most effective means of understanding and improving the techniques for measurement. Extraneous inputs still include micrometer errors due to temperature, although the environment in most laboratories is well enough controlled to eliminate this source. The machining error is still present.

INFEROMETER–MONOCHROMETER

The wave characteristic of light has become important in the concepts of measuring length. The present standard for length measurement (the wavelength of a krypton light source) is based on the wave nature of light. We are all relatively familiar with the alternating electrical voltage present in our homes, voltage $= V_0 \sin \omega t$. For house current $V_0 = 115$ V and $\omega = 60$ Hz. In Figure 6.9 the observer at location A in the wire measures

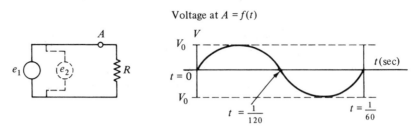

Figure 6.9 The Wave Characteristic of Alternating Voltage

a voltage of zero at $t = 0$, a voltage of V_0 at $t = \frac{1}{240}$ s, a voltage of zero at $t = \frac{1}{120}$ s, a voltage of $-V_0$ at $t = \frac{1}{80}$ s, and a voltage of zero at $t = \frac{1}{60}$ s. If a second voltage source, e_2, is introduced, with $e_2 = V_0 \sin(120\pi t + \pi)$, the voltage observed at A is zero. The voltages are out of phase by $\frac{1}{2}$ wavelength $(\lambda/2 = 2\pi/2 = \pi)$. The voltage wave passes through the wire with a speed of approximately the speed of light, and the length of wire from e_1 to A must be equal to the length of wire from e_2 to A, or there will be an additional phase shift due to wave propagation.

Electrical voltage is at one end of the radiation spectrum, with frequency at approximately $120\pi/s$. The total radiation spectrum includes all frequencies and is represented by

$$E = \sum_{i=1}^{\infty} E_i \sin 2\pi \omega_i t = \sum_{i=1}^{\infty} E_i \sin \nu_i t$$

The wavelength of light is defined to be the ratio of the velocity of propagation in a given medium to the frequency of the wave, $\lambda_i = V_p/\nu_i$. A wedge of material capable of transmitting (allowing light to pass through) light can be used to divide white light (light with all frequencies, or wavelengths, present) into the individual wavelengths because of the interference nature of light waves. Figure 6.10 shows the separation of wavelengths in the visible range. The incoming light at point 1 contains all wavelengths (or fre-

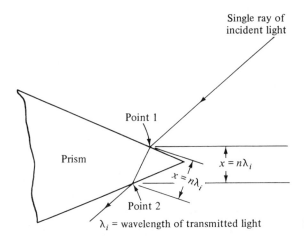

Single ray of
incident light

Point 1

Prism

$x = n\lambda_i$

$x = n\lambda_i$

Point 2

λ_i = wavelength of transmitted light

Figure 6.10 Prism Separation of Light

quencies), but at point 2 only the light with wavelength equal to an integer, n, divided into the propagation length, x, will be in phase. All other wavelengths will be out of phase and will exhibit destructive interference. Since the propagation length, x, varies with the position along the wedge, the wavelengths transmitted will vary along the wedge. You probably observed this in early childhood by dividing "white light" into the colors of the rainbow, using either a prism or the earth's atmosphere.

A monochrometer is identical to the prism, but it also includes a movable slot at point 2 to allow the net output of light waves to be essentially composed of a very small range of wavelengths. More sophisticated monochrometers employ a double pass through the prism to further displace the wavelengths. The monochrometer is able to provide light with a very small band of wavelengths. This is frequently called a single-wavelength light source, although some error limits are imposed for a wavelength range.

A second source available for monochromatic (single-wavelength) light is the laser. However, each laser is able to provide only light having a fixed wavelength, while the monochrometer is capable of providing light having a range of wavelengths. The laser has one additional advantage in that the light wave produced is already highly collimated and does not generally require focusing lenses. The laser also produces some wavelengths in the side bands of the central wavelength. However, it is more precise than the monochrometer.

The interferometer operates on the principles already discussed. Figure 6.11 shows a simple schematic diagram of the interferometer. A single wavelength of light is emitted from the source, and a beam splitter allows one-half of the light to pass to the object mirror and reflects the other half

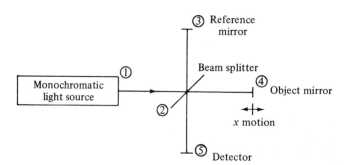

Figure 6.11 Interferometer Schematic Diagram

to the reference mirror. Each mirror then reflects the light back to the beam splitter to the detector. With the object to be measured at position 4, fixed, the paths 1–2–3–2–5 and 1–2–4–2–5 are adjusted to have the same effective lengths by positioning the reference mirror, 3 (L_1). The position is correct when the detector senses constructive (or additive) interference of the single-wavelength light at point 5. If the object mirror moves so that $x = \lambda/4$, the path length 1–2–4–2–5 increases by an amount $\lambda/2$. The detector will observe destructive interference, so the intensity of light received is zero. If the wavelength is known, the value of the length moved, x, is measured.

If the light source contains several different wavelengths closely grouped around the known frequency, the interference will be not as distinct as for the single-wavelength source, and an extraneous error will result from the fringe definition. The methods discussed here are employed with a krypton light source to provide calibration of length measurement with an accuracy of $\pm(\frac{1}{4}$ wavelength)(1 in./41,929.399 wavelengths) $\doteq \pm 6 \times 10^{-6}$ in. (15.24 × 10^{-6} cm). There are other techniques for separating the fringe patterns to obtain a wavelength measurement more precise than $\pm\frac{1}{4}$ wavelength, but these will not be necessary for the present introduction.

A monochromatic light source can be used to produce a fringe pattern for length measurement. Figure 6.12 shows one possible application. When the path length from the beam splitter to the mirror and back to the beam splitter is an integral multiple of wavelength, constructive interference occurs and a light spot appears on the fringe pattern. This makes a pattern of light and dark lines. Counting the number of light spots and knowing the wavelength allows one to measure the unknown length.

PHOTOGRAPHS

The use of photography in the measurement of length actually introduces no new measurement concepts. The photograph is only a tool for transfer-

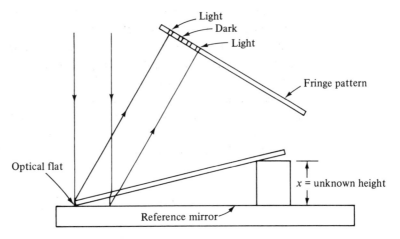

Figure 6.12 Fringe Patttern for Measurement

ring measurement information to a permanent record. The use of high-speed photography has introduced the need for clarifying the basic principles of length measurement. Basically, the problem is to include on the photograph a grid system of known lengths in the focal plane of the camera object. Preferably, the scale or grid should be placed alongside the object to be measured. If this is not possible, camera optics must be known and the theories of optics must be applied to correct for focal length, depth of field, and projection techniques. The use of slides and projections can magnify or amplify the scale to very wide ranges. The major limitation on this amplification is the clearity of the original slide. Photography is also used in conjunction with interferometers to provide a permanent record fringe patterns. Again, magnification allows for a more precise determination of the number of wavelengths between a reference dimension and an unknown dimension (Figure 6.12).

ELECTRICAL SYSTEMS FOR LENGTH MEASUREMENT

The methods available for measuring electrical voltages and current will be developed in this chapter. Although length is a fundamental dimension, the use of electrical sensors in measurement systems has provided the technical capabilities for transducing a linear dimension into a corresponding electrical signal. The advantages of electrical systems are principally that they are inertialess, frictionless, easy to amplify, easy to transmit, and that sensors are small in size. Because many measurement systems employ electrical sensors, it is appropriate to introduce the systems frequently used in the measurement of length at this time. Since most electrical properties do depend on length, only the most commonly used sensors will be discussed in detail. However, any electrical property that can be expressed as a function of length could be used to sense length.

It is also necessary to introduce two basic electrical circuits and some concepts of these circuits. The ballast circuit and the Wheatstone bridge circuit are almost always used in conjunction with sensors of electrical resistance. The ballast circuit of Figure 6.13 is analyzed by writing the volt-

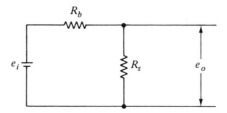

Figure 6.13 Ballast Circuit

age loop equation for the circuit,

$$e_i = i(R_b + R_s) \tag{6.1}$$

where the exciting voltage is e_i, the output voltage is e_o, the current flow is i, the ballast resistor is R_b, and the sensor resistance is R_s.

$$e_o = iR_s \tag{6.2}$$

Then

$$\frac{e_o}{e_i} = \frac{R_s}{R_b + R_s} \tag{6.3}$$

The ballast resistor should be selected to provide the maximum output-voltage change for a change in the sensor resistance. The sensitivity of any measurement system is defined to be the change in output information with respect to the input information,

$$\eta = \frac{dI_o}{dI_i} \tag{6.4}$$

For the ballast circuit,

$$\eta = \frac{de_o}{dR_s} = \frac{-e_i R_b}{(R_b + R_s)^2} \tag{6.5}$$

The optimal values of ballast resistor to optimize sensitivity can be found by requiring that

$$\frac{d\eta}{dR_b} = 0 = \frac{-e_i}{(R_b + R_s)^2} + \frac{2e_iR_b}{(R_b + R_s)^3} = \frac{e_i}{(R_b + R_s)^3}(R_b - R_s) \quad (6.6)$$

There are two values of R_b which satisfy Equation 6.6, $R_b = \infty$ and $R_b = R_s$. The first condition gives a sensitivity, from Equation 6.5, equal to zero. Also, it can be observed that when $R_b = 0$, the sensitivity is also zero. However, when $R_b = R_s$, the sensitivity is optimal (from Equation 6.6), and it is not zero; therefore, it is a maximum. Practically, the sensor resistance, R_s, changes over a range of values, and R_b should be selected to be the average sensor resistance in the range where the system is to operate. It should also be stated that purely resistive circuits are zeroth-order elements in a measurement system (this statement neglects the time required for a voltage to propagate the length of the wire, $t = L/V_p$). If a direct current is used to excite the circuit, capacitance and inductance effects are essentially zero (neglecting noise).

The Wheatstone bridge circuit is shown in Figure 6.14. The loop equations

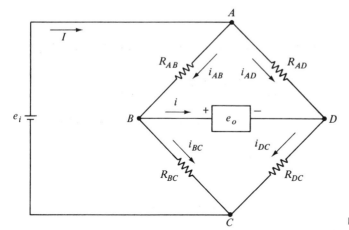

Figure 6.14 Wheatstone Bridge Circuit

can be written around any closed loop to give, for voltages

(1) $ABCe_i$ $e_i = i_{AB}R_{AB} + i_{BC}R_{BC}$ (6.7)

(2) $ADCe_i$ $e_i = i_{AD}R_{AD} + i_{DC}R_{DC}$ (6.8)

(3) ABD $i_{AB}R_{AB} + e_o = i_{AD}R_{AD}$ (6.9)

(4) BDC $e_o + i_{DC}R_{CD} = i_{BC}R_{BC}$ (6.10)

(5) $ABCD$ $i_{AB}R_{AB} + i_{BC}R_{BC} = i_{DC}R_{CD} + i_{AD}R_{AD}$

Also Kirchhoff's law for current flow to a junction gives

(1) $I = i_{AB} + i_{AD}$
(2) $i_{AB} = i + i_{BC}$
(3) $i_{BC} + i_{DC} = I$
(4) $i + i_{AD} = i_{DC}$

If the null bridge is used, the current flow through the output meter is zero. (This is also essentially true when a voltage-balancing circuit is used to measure output voltage or when a very-high-resistance voltage-measuring device is used. Vacuum-tube voltmeters and oscilloscopes have sufficiently high resistance to satisfy this condition.) For the nullbridge, the voltage at B is equal to the voltage at D. In both cases, the current through the output meter is zero. Then

$$i_{AB} = i_{BC} \quad \text{and} \quad i_{AD} = i_{DC} \tag{6.11}$$

The loop equations become

$$e_i = i_{AB}(R_{AB} + R_{BC}) \tag{6.12}$$
$$= i_{AD}(R_{AD} + R_{AC}) \tag{6.13}$$

If the output voltage is zero, $e_o = 0$ and Equations 6.9 and 6.10 give

$$\frac{i_{AB}}{i_{AD}} = \frac{R_{AD}}{R_{AB}} \quad \text{and} \quad \frac{i_{BC}}{i_{DC}} = \frac{R_{DC}}{R_{BC}} \tag{6.14}$$

Using Equations 6.11 and 6.14, the condition for null balance is

$$\frac{i_{AB}}{i_{AD}} = \frac{i_{BC}}{i_{DC}} = \frac{R_{AD}}{R_{AB}} = \frac{R_{DC}}{R_{BC}} \tag{6.15}$$

or

$$R_{AD}R_{BC} = R_{AB}R_{DC} \tag{6.16}$$

The products of the resistances in opposite legs at a Wheatstone bridge are equal when the output voltage is zero. If the output meter has infinite resistance so that no current flows through that circuit, the equations for deflection-bridge operation ($e_o = 0$) can be written. From Equations 6.7 and 6.8,

$$i_{AB} = i_{BC} = \frac{e_i}{R_{AB} + R_{BC}} \tag{6.17}$$

$$i_{AD} = i_{DC} = \frac{e_i}{R_{AD} + R_{DC}} \tag{6.18}$$

and the output voltage can be determined from Equations 6.9 and 6.10,

$$e_o = i_{AD}R_{AD} - i_{AB}R_{AB} \qquad (6.19)$$

$$= i_{BC}R_{BC} - i_{DC}R_{CD} \qquad (6.20)$$

Combining the last four equations gives

$$\frac{e_o}{e_i} = \frac{R_{AD}}{R_{AD} + R_{DC}} - \frac{R_{AB}}{R_{AB} + R_{BC}} = \frac{R_{BC}}{R_{AB} + R_{BC}} - \frac{R_{CD}}{R_{CD} + R_{AD}} \qquad (6.21)$$

Obtaining a common denominator gives

$$\frac{e_o}{e_i} = \frac{R_{AD}R_{BC} - R_{AB}R_{CD}}{(R_{AB} + R_{BC})(R_{AD} + R_{CD})} \qquad (6.22)$$

Each resistor is called a leg of the Wheatstone bridge. A leg (resistor) is said to be an active leg if the resistance changes during the operation. An equivalent form of Equation 6.22 results if the output voltage, e_o, is assumed to have the opposite sign of that in Figure 6.14:

$$\frac{e_o}{e_i} = \frac{R_{AB}R_{CD} - R_{AD}R_{BC}}{(R_{AB} + R_{BC})(R_{AD} + R_{DC})} \qquad (6.23)$$

Figure 6.15 is another commonly used system for specifying the resistance.

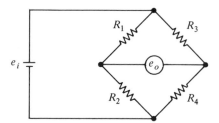

Figure 6.15 Wheatstone Bridge Circuit

For this system Equation 6.23 becomes

$$\frac{e_o}{e_i} = \frac{R_1 R_4 - R_2 R_3}{(R_1 + R_2)(R_3 + R_4)} \qquad (6.24)$$

If R_1 is the only active leg, the input resistance is R_1 and the sensitivity is

$$\eta = \frac{de_o}{dR_1} = e_i \frac{R_2 R_4 + R_2 R_3}{(R_1 + R_2)^2(R_3 + R_4)} \qquad (6.25)$$

To select the optimal values for fixed resistors R_2, R_3, and R_4, the partial derivative of sensitivity with respect to each resistance is set equal to zero.

$$\frac{\partial \eta}{\partial R_2} = 0 = e_i \frac{R_4 + R_3}{(R_1 + R_2)^2(R_2 + R_3)} \frac{2(R_2 R_4 + R_2 R_3)}{(R_1 + R_2)^3(R_3 + R_4)}$$

$$= \frac{R_1 R_4 + R_1 R_3 + R_2 R_4 + R_2 R_3 - 2R_2 R_4 - 2R_2 R_3}{(R_1 + R_2)^3(R_3 + R_4)}$$

$$= \frac{R_1(R_3 + R_4) - R_2(R_3 + R_4)}{(R_1 + R_2)^3(R_3 + R_4)} \tag{6.26}$$

This last expression is zero when $R_2 = \infty$ (not a maximum) and when $R_2 = R_1$ (the maximum value).

$$\frac{\partial \eta}{\partial R_3} = 0 = e_i \frac{R_2}{(R_1 + R_2)^2(R_3 + R_4)} - \frac{(R_2 R_3 + R_2 R_4)}{(R_1 + R_2)^2(R_2 + R_4)^2}$$

$$= \frac{R_2 R_3 + R_2 R_4 - R_2 R_3 - R_2 R_4}{(R_1 + R_2)^2(R_3 + R_4)^2}$$

This is zero for all values of R_3 and R_4. It is interesting to note that when $R_3 = \infty$ and $R_4 = 0$, the Wheatstone bridge circuit is identical to the ballast circuit. The partial of η with respect to R_4 gives the same result as that for R_3. Therefore, the maximum sensitivity for one active leg occurs when $R_1 = R_2$.

When all four legs are active, the output voltage is

$$\frac{e_o + \Delta e_o}{e_i} = \frac{(R_1 + \Delta R_1)(R_4 + \Delta R_4) - (R_2 + \Delta R_2)(R_3 + \Delta R_3)}{(R_1 + \Delta R_1 + R_2 + \Delta R_2)(R_3 + \Delta R_3 + R_4 + \Delta R_4)} \tag{6.27}$$

If the bridge is initially balanced (null condition), $R_1 R_4 = R_2 R_3$, then $e_o = 0$ and Equation 6.27 becomes

$$\frac{\Delta e_o}{e_i} = \frac{(R_1 \, \Delta R_4) + (R_4 \, \Delta R_1) + (\Delta R_1 \, \Delta R_4)}{(R_1 + \Delta R_1 + R_2 + \Delta R_2)(R_3 + \Delta R_3 + R_4 + \Delta R_4)} \tag{6.28}$$

If, additionally, $R_1 = R_2 = R_3 = R_4$ initially and higher-order terms are neglected in comparison to the initial values (i.e., $\Delta R, \Delta R_4, \ldots$ are neglected). Equation 6.28 becomes

$$\frac{\Delta e_o}{e_i} = \frac{R_1 \, \Delta R_4 + R_4 \, \Delta R_1 - R_2 \, \Delta R_3 - R_3 \, \Delta R_2}{(R_1 + R_2)(R_3 + R_4)}$$

$$= \frac{\Delta R_4 + \Delta R_1 - \Delta R_2 - \Delta R_3}{4R_1} \tag{6.29}$$

Equation 6.29 is frequently useful in strain-gage circuits.

To obtain a null balance, initially the circuits of Figure 6.16 may be used. Other methods are available but will not be discussed here. For precision bridge circuits, the length of wire from points A and C of Figure 6.14 to the power supply, e_i, should be the same and the wire identical. Best results are obtained when all joints are soldered and all metal is shielded from noise. Now it is possible to return our attention to electrical sensors for length measurement.

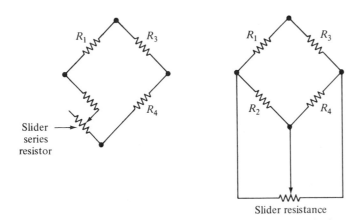

(a) Series balance   (b) Differential balance

Figure 6.16 Balancing Circuits for Wheatstone Bridge

RESISTANCE POTENTIOMETERS

The slider-type resistance potentiometer is made in the two basic systems of Figure 6.17. When the slider passes over a continuous rod, the resistance varies linearly with distance. However, when the slider crosses coils of wire, the resistance may change its steps. Dirt and oil on the slider must be avoided. The principle of operation of the resistance potentiometer is the equation for electrical resistance,

$$R = \frac{\rho L}{A} \tag{6.30}$$

where R is the resistance, ρ is the resistivity of the resistance, L is the resistance length, and A is the cross-sectional area of the resistance element. The sensor is also sensitive to temperature, strain, humidity for some elements, and electrical contact resistance. This latter effect can be extremely severe on occasions. The system is simple, inexpensive, can have high

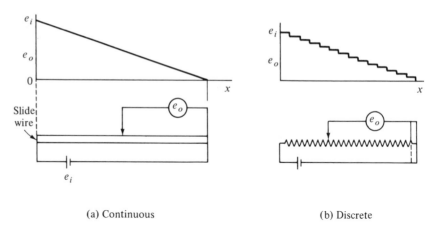

(a) Continuous (b) Discrete

Figure 6.17 Resistance Potentiometers

response, requires simple modifying and read-out devices, and can have high sensitivity and wide range by proper construction.

Resistance potentiometers can be spring loaded to maintain contact with a moving object. They can be constructed around a cylinder to obtain 10–20 revolutions of range. Their simplicity and precision make them ideal systems for static measurement. However, at high frequencies the slider may actually lose contact with the resistance element.

Strain gages are also governed by Equation 6.30, but their use for measuring length will be deferred until Chapter 7. The special techniques for application and the associated yielding and fatigue problems will be discussed in detail.

LINEAR VARIABLE-DIFFERENTIAL TRANSFORMERS

The linear variable-differential transformer (LVDT) is shown in Figure 6.18. The primary (or input) voltage excites the core, made of a magnetic material. The mutual inductance in the secondary (output) coils will exactly cancel each other when the core is in the center position, because the secondary coils are wound in phase opposition. The LVDT is linear over a limited range. An output detector capable of sensing phase could differentiate between $+x$ and $-x$ displacement. However, a voltmeter would give only a magnitude of displacement and no direction. Both exciting and output voltages are ac, and the frequency of the exciting voltage should be 10 times the frequency of motion of the core to be detected for dynamic systems.

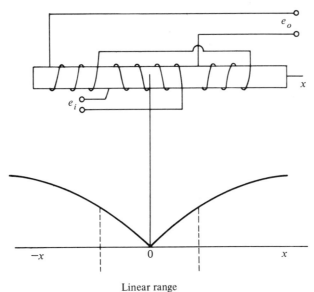

Figure 6.18 Linear Variable Differential Transformer

Linear range

If the LVDT is to operate in the proximity of any magnetic material, the electrical and magnetic fields will be disturbed. Inductance is a function of the number of coils in the windings, the length of core in the field, the permeability of the core material, the cross-sectional area of the core and air gap, and the length of the air gap:

$$L = \frac{\text{const. (no. turns)}^2}{(L_{\text{iron}}/\mu_i A_i) + L_{\text{air}}/A_{\text{air}}}$$

Precision LVDT's are relatively expensive because of the knowledge and workmanship required to make them. However, they can have high response. They do not require complicated modifying and read-out devices, and they are generally capable of precision measurement. Extraneous inputs of electrical noise, humidity, and temperature must be considered. Certainly the LVDT does not involve the electrical contact problem present in the slide-wire potentiometer.

VARIABLE-CAPACITANCE DEVICES

The capacitance of two parallel plates depends on the plate area, the dielectric of the media separating the plates, and the distance between the plates:

$$C = \frac{(const)(dielectric\ constant)(area)}{distance}$$

Two methods of changing capacitance with changing length are shown in Figure 6.19. For a small range of distances the capacitance varies with plate motion [part (a)]. If the plates are moved up and down in a non-

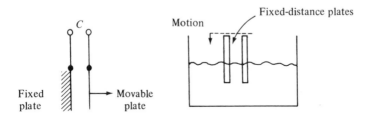

(a) Plate motion   (b) Variable dielectric constant

Figure 6.19 Variable Capacitance Devices

conducting fluid, the effective dielectric constant varies with vertical motion. In part (a) the range is limited, and in both cases the modifying elements are much more complex than for the previous sensors. Many laboratories do not have secondary standards for the calibration of capacitance. However, an ac Wheatstone bridge circuit with two fixed capacitors and one variable capacitor could be used. Humidity and temperature cause variation in the dielectric constant, and electrical noise can be a problem with this type of sensor.

The piezoelectric crystal and the magnetic pick-up will be considered later. They actually perform primarily under dynamic conditions to measure velocity. Velocity-measuring systems are considered in Chapter 7.

6.3
Measurement Systems for Mass

The primary standard for mass is a platinum–iridium mass maintained at Sèvres, France, and is defined to be 1 kg. The NBS maintains a secondary standard mass. In 1959 the avoirdupois pound was redefined in terms of grams:

$$1\ lb\ avoirdupois = 453.59237\ g$$

Mass standards can be certified by the NBS for use as working laboratory standards.

The basic measurement system for comparing masses is the beam balance. Figure 6.20 shows the essential components of the beam balance. For precision measurements four considerations should be made. First,

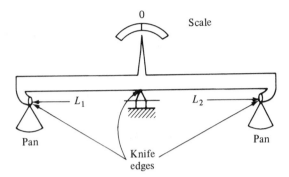

Figure 6.20 Beam Balance

the entire measurement system should be placed in an evacuated chamber, or the buoyancy of the air must be considered when the density of the unknown mass is not equal to the density of the standard mass.

$$\text{buoyancy of air} = V_{\text{mass}}\rho_{\text{air}}$$

$$= \frac{M_{\text{ass}}}{\rho_{\text{mass}}}\rho_{\text{air}}$$

For balance at zero with equal lever arms, L_1 and L_2, the unknown mass less the buoyancy equals the standard mass less its buoyancy:

$$M_u - B_u = M_{\text{st}}B_{\text{st}}$$

$$M_u - \frac{M_u\rho_{\text{air}}}{\rho_u} = M_{\text{st}} - \frac{M_{\text{st}}\rho_{\text{air}}}{\rho_{\text{st}}}$$

$$M_u = M_{\text{st}}\frac{1 - \rho_{\text{air}}/\rho_{\text{st}}}{1 - \rho_{\text{air}}/\rho_u}$$

Second, the knife edges must be sharp and accurately located. If the knife edges are of measurable width, the system lengths are uncertain. Then,

$$M_u = M_{\text{st}}\frac{1 - \rho_{\text{air}}/\rho_{\text{st}}}{1 - \rho_{\text{air}}/\rho_u}\frac{L_{\text{st}}}{L_u}$$

Third, the system must be level with the horizontal. If this condition is not met, only null operation is possible. It is also desirable to place the center of mass of each mass in the center of the pan. This helps avoid unwanted couples. Finally, the beam should be balanced with no masses on the pans. A calibration using two standard masses is also worthwhile in the measurement technique.

Other physical mass-balancing systems employ the same principle as the beam balance ($\sum$ moments $= 0$). These systems are for the measurement of static mass only. The beam balance is a second-order system with very small damping. It will oscillate for long periods before coming to rest. In practical measurement the operator introduces friction by stopping the indicator at zero and holding it there for a few seconds before releasing it. The platform scale is a composite of levers and knife edges, but it actually obeys the same law that governs the beam balance.

Air currents, friction, and knife-edge width are extraneous inputs. For heavy loading the effect of stress and strain could become important. Any dirt on the standard masses could cause error. Also, the standard masses wear with prolonged use (or by dropping one accidentally). It should be noticed that the local acceleration of gravity in no way influences the measurement of mass with the beam balance.

Other methods of measuring mass are available (since $m = V\rho$, knowing density and measuring volume would determine the mass). However, the particular method chosen to measure mass would employ the measurement of some system not yet discussed. Therefore, the mass-measurement system suggested at this point is the beam balance or some more refined beam balance.

6.4
Time- and Frequency- Measuring Systems

Time is the most nebulous dimension present in physical measurement. It has been established based only on an astronomical phenomenon, and it is the one completely independent dimension. Philosophically, it cannot be separated from each individual life; and the path of time is, apparently, irreversible. To our knowledge, no event has been able to move backward in time. Only history records the events of time. Equally, future time has not been explored. Therefore, time is the present, and scientific exploration is involved with the history of the past and a prediction of the future.

Fortunately, scientific technology is concerned with elapsed time; and the methods for measuring this quantity far exceed the requirements for the most general problems. Historically, time has been measured by astronomical events and is limited by astronomical measurements. The number of moons or suns passing has allowed a measure of elapsed time. Typically, the time required for the earth to assume the same position with respect to the sun has been called 1 year. This time has been divided into days (suns) to establish the mean solar day. The second was originally defined to be the time required to accomplish 1/86,400 of a revolution of the earth on its axis on a mean solar day. Since the earth was discovered to be slowing down at the rate of 0.001 s/century, the second could be a standard if it were based on the revolutions for one particular year. In 1956 the second was defined to be the time required for the earth to complete 1/31,556,925.9747 of a revolution around the sun in the year 1900.

The error in astronomical observations could not properly assess the precision error in the foregoing definition. The invention of the atomic clock led to the definition of the second in 1967 as the time required for 9,192,631,770 periods of radiation in the transition between two levels of the cesium-133 atom (see reference 1). This definition allowed the NBS to disseminate frequency standards from Boulder, Colorado, with an accuracy of one part in 10^8. Station WWV from Fort Collins, Colorado, broadcasts standard frequency standards of 600, 500, and 440 Hz on radio frequencies of 2.5, 5, 10, 15, 20, and 25 MHz with an accuracy of one part in 10^8. These services provided are influenced by ionospheric conditions. The ionospheric errors can be as large as 1 μs.

The basic standard for time-frequency systems is station WWV in Fort Collins, Colorado. The basic measurement systems vary from mechanical to electrical to atomic.

MECHANICAL SYSTEMS

The basic mechanical systems available for the measurement of time (or frequency) are the pendulum, the spring mass, and the tuning fork. The pendulum became one of the first time-measuring devices when it was discovered that the oscillations of the pendulum were constant for a given system. The "grandfather clock" could record time digitally, and it was capable of dividing the second into smaller parts. The pendulum can be analyzed from Figure 6.21. The mass times the tangential acceleration

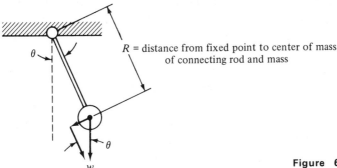

R = distance from fixed point to center of mass
of connecting rod and mass

Figure 6.21 The Simple Pendulum

is equal to the restoring force:

$$\frac{w}{g} R \frac{d^2\theta}{dt^2} = -w \sin \theta \doteq -w\theta \tag{6.31}$$

This is true for small angular displacements. Then

$$\frac{d^2\theta}{dt^2} + \frac{g}{R}\theta = 0 \tag{6.32}$$

And with a start at $\theta = \theta_0$, $d\theta/dt = 0$,

$$\theta = \theta_0 \cos \frac{g}{R} t \tag{6.33}$$

The frequency of oscillation is

$$f = \frac{1}{2\pi}\sqrt{\frac{g}{R}} \text{ Hz} \tag{6.34}$$

The frequency changes with temperature and local gravity; the damping due to air friction is eliminated by providing an energy source to balance the dissipation. Therefore, the pendulum mass and energy source can be used to provide a relatively constant frequency source. With proper linkages this frequency is converted into elapsed time.

The mainspring in a precision watch operates on the same principle. (A strain gage on a vibrating beam is also the same.) The restoring force depends on the spring constant instead of gravity. The air friction and external disturbances can be offset by the storage of energy in an additional storage spring. The frequency of vibration of the mainspring and mass can be altered by adjusting the fixed end of the circular-wound mainspring. Temperature is still an extraneous input to the watch. It is possible to obtain watches capable of measuring time with an accuracy of ± 0.01 s. Theoretically, the accuracy could be greater than this using the spring–mass system; but practically, the capabilities of vibrating crystals are more satisfactory for operation when accuracies greater than ± 0.01 s are required.

The tuning fork, when excited, has a resonant vibrational frequency which depends on the physical dimensions and the material of the fork. Air-pressure waves propagating from the fork can also be used for known signal sources. If one side of the fork is excited with a signal that is an even multiple of the resonant frequency, the other side will provide a mechanical motion that has the basic frequency of the fork. This output motion can be used to provide a frequency in the higher ranges. The output frequency is influenced by temperature and the medium through which the pressure waves are propagated. The damping effect of air is very small, and the resonant frequency can be maintained for relatively long periods in an air medium.

The mechanical vibrations of a piezoelectric crystal excited by pressure pulses can provide a periodic voltage output. This particular sensor will be discussed in Chapter 7. However, a crystal oscillator excited by a source signal is used to provide the oscillating frequency standard that is the principal element in most electronic systems.

ELECTRONIC COUNTERS AND ELEMENTS

The basic elements of the electronic counter are given in Figure 6.22. The particular element capable of measuring elapsed time is the internal oscillator. The oscillator frequency is influenced primarily by temperature,

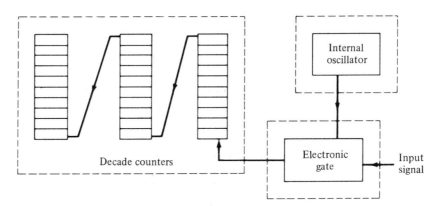

Figure 6.22 Elements of an Electronic Counter

and precision counters are provided with a constant-temperature oven environment to house the crystal oscillator. The decade counters may number up to approximately nine with an error of plus or minus one count in nine significant figures. The counter accuracy must be checked periodically by applying a known signal from an atomic clock, if possible. However, five to six significant-figure calibrations can be accomplished using the broadcast from station WWV.

The counter operates in the three modes previously discussed. For calibration purposes, the EPUT (events per unit time) mode may be selected and the counter can count the basic broadcast frequencies from station

WWV. A typical experimental procedure might be to compare the Lissajous patterns, comparing the frequency broadcast from WWV to that of a variable-frequency oscillator or an oscilloscope. The oscilloscope acts as an x-y, plotter; the x input is from WWV and the y input is from the variable-frequency oscillator. The variable frequency is adjusted until the Lissajous pattern is a fixed or standing pattern. For example, when the broadcast frequency is 600 Hz, $x = A \sin 1200\pi t$, and a standing wave results when the variable frequency is adjusted so that the ratio of the broadcast frequency to the variable frequency is reducible to a ratio of integers. It is generally possible to interpret the patterns when the ratio is less than 10 to 1. With very good reception the following patterns 10–9, 5–4, 10–7, 5–3, 2–1, 5–2, 10–3, 5–1, 10–1, 9–8, 9–7, 3–2, 9–5, and so on can be observed. The inverse patterns (1–10, 2–3) are also possible. With $x = A \sin 1200\pi t$ and $y = 600\pi t$, the Lissajous pattern of Figure 6.23 would occur. When variable

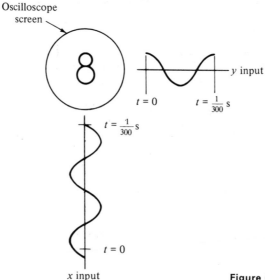

Oscilloscope screen

$t = 0$ $t = \dfrac{1}{300}$ s y input

$t = \dfrac{1}{300}$ s

$t = 0$

x input

Figure 6.23 Lissajous Pattern (2 to 1)

frequency is slightly different from 300 Hz, the pattern will rotate on the screen. When the frequency is exactly 300 Hz, the pattern will be absolutely fixed. For the Lissajous patterns given, station WWV can be used to obtain calibration points for the variable frequency oscillator from 44 to 6000 Hz.

Once the pattern on the scope is fixed and recorded, simultaneous counting of the signal from the oscillator by the electronic counter is theoretically the same as counting the WWV broadcast. The advantage of this method lies in the fact that the WWV broadcast usually has a considerable amount of noise. When the Lissajous pattern does not rotate (even though it still has noise jitters), the stable signal from the oscillator is a known multiple of the WWV signal. The same concept can be extended with the addition of another oscillator and oscilloscope. Then the signal from the first counter can be used to obtain 10–1 or 1–10 multiples of the second oscillator on the second scope. The signal from the second oscillator will be from 4.4 to 60,000 Hz. The work involved and the time limitation of the broadcast signal will introduce additional sources of possible error in the calibration.

When the counter operates in the PER (period) mode, the same type of signal from the first oscillator can be used to evaluate the oscillating frequency of the internal oscillator of the counter. With these two checks (or with more refined calibration equipment), the counter can be certified for laboratory use within defined limits of accuracy. The calibration should be performed over the entire range of possible time period selections in the EPUT and over the range of periods in the PER mode (usually 1 or 10 periods are selected).

Three other control selections are possible in the counter operation: the triggering level, the triggering slope, and two separate input signals for start–stop operation. The triggering level–slope operation is discussed with respect to Figure 6.24. If the voltage level of the input signal is above the

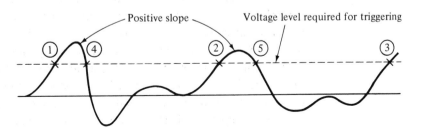

Figure 6.24 Triggering Level-Slope for Counters

selected triggering level and the slope of the input signal is positive, the input signal is counted at points 1, 2, and 3 if the mode of operation is EPUT (this assumes that the time elapsed between points 1 and 3 is less than the time interval selected). For PER operation, the gate will open on 1, close on 2, and open again on 3. If the negative slope is chosen, the counts occur at 4 and 5 for the same triggering level. Notice that the lower voltage peaks are not counted. Proper selection of triggering level can be used to eliminate noise from the signal.

If one signal is to start the counting of the signal from the internal oscillator and a second signal is to stop the counts, the counter can be used to determine the phase shift between two signals. The counter simply counts the time elapsed between the start signal and the stop signal. The timing of a 100-m race can be accomplished by pushing a button which fires a gun and starts the counter. The counting will stop when the first racer breaks a beam of light directed to a photomultiplier tube.

It is also possible to control the display time on the decade counters. This will allow time for recording the results before another event is started. This additional control locks the electronic gate after an operation has been completed. The time required is variable and is controlled by the selection of display time.

The counter can be influenced by temperature, electrical noise, and electronic gate operation. Very accurate calibration is possible, and the significance of the count can be evaluated using steady signals. The most accurate mode of operation is the mode that provides the largest number of significant figures. (This is almost always the number of decades used ± 1 count in the least decade counter.)

The variable-frequency oscillator and the 60-Hz line voltage provided by electrical power companies are two other frequency sources available for measuring time. The oscillator was discussed in the calibration of the electronic counter. It should be noted that the same test also calibrates the variable-frequency oscillator for each Lissajous pattern established. These are all point calibrations, and additional scale divisions must be used to define the frequencies between the calibration points. The output voltage of the oscillator may be a sine wave, square wave, triangular wave, or any other periodic wave form. The basic calibration of these electronic devices establishes them for use as secondary or working standards. They may be used to calibrate oscillographs, oscilloscopes, magnetic tapes, and so on for the measurement of time. The 60-Hz line voltage provided by power companies may have errors at any specific time of the day, but the total number of cycles generated for one 24-h period is adjusted (usually at midnight) so that $(24 \text{ h})(3600 \text{ s/h})(60 \text{ c/s}) = 5,184,000$ cycles are generated each day. This provides electrical clocks with the correct time-keeping capabilities.

Every physical phenomenon is influenced in some way by time and is, therefore, capable of measuring time. However, the extraneous inputs and degree of influence make most of them incapable of accurate time measurement. Granite rock erodes with time, but the erosion is strongly influenced by rain, ice, collisions with other rocks, and man. The evaluation of time using a granite-rock sensor would be very inaccurate compared to a rotation of the earth on its axis. If the distance between two points is known and the time is known when a pulse of light was generated, the speed of light could be used to determine time. (The medium through which it was propagated influences the speed of light.) Additional physical phenomena are available for the measurement of time, and any one might prove to be entirely adequate in a given situation (i.e., counting 1000–1, 1000–2, 1000–3 gives a rough measure of 3 s elapsed time to experienced counters). The systems discussed in detail in this section are the basic time-measuring systems on which future measurement of time in this book will be based. The reader is reminded that the accuracy required for a particular experiment dictates the type of measurement system to be selected. The geologist may use the half-life of elements in his time measurement, while the space scientist may require an atomic clock.

6.5 Temperature-Measuring Systems

The dimension temperature can be derived from the basic dimensions previously defined. In order for a defined standard temperature scale to satisfy the physical laws, it must agree with the previous dimensions. One basic temperature scale is derived from thermodynamic considerations. This definition is fundamentally involved with the Carnot cycle of thermodynamics. Although time, mass, and length are well defined, their interrelationship with temperature is a complex concept. Therefore, the temperature scales have been defined by establishing fixed temperature points and interpolating between these points. The arbitrary choice of one temperature (the triple point of water) completely specifies the temperature scale. The commonly used Celsius scale was developed by assigning the triple point of water a value of 0° and the boiling point of water a value of 100°. The equivalent values selected for the Fahrenheit scale are 32° for the triple point and 212° for the steam point. This type of definition requires experimental determination of an absolute zero of temperature.

The ideal gas thermometer allows the theoretical realization of the thermodynamic temperature scale. However, all gases are real and considerations for this fact must be made before the ideal gas thermometer can be used. The basic temperature scale has been defined at fixed points by easily reproduced thermodynamic states. The thermodynamic scale assumes a reversible process, which has never been realized for a real measurement system. It also assumes that the entire system is in thermodynamic equilibrium. This also cannot be absolutely realized (although some small volume may approximate this very closely). Therefore, this ideal temperature scale is fixed by agreement with the best accuracy available.

The international Practical Temperature Scale of 1948, with revision in 1960, has been adopted (see reference 2). This scale was defined with six reproducible temperature points and methods for interpolating between the points. The value assigned to each point was to conform to the thermodynamic temperature scale with the maximum known accuracy available at that time. When additional knowledge concerning deviations from the thermodynamic scale are known, the defined points will be changed to conform. From thermodynamics, the temperature of a homogeneous, single-component, simple-compressible substance in the two-phase region is dependent on the pressure. All equilibrium points except the triple point of water have been established on the International Practical Temperature Scale at a pressure of 101,325 N/m² (14.6959 psia). Table 6.1 gives the fixed-point temperature standards presently used.

The interpolation between various fixed-point standards is recommended using a platinum resistance thermometer and a platinum and platinum–rhodium alloy thermocouple.

A. From the oxygen point to the ice point,

$$R_t = R_0[1 + At + Bt^2 + C(t - t_{100})t^3] \qquad (6.35)$$

where R_t is the resistance of the thermometer at temperature t in degrees Kelvin, R_0 is the resistance of the thermometer at 0°C, $t_{100} = 100°C$, and A, B, and C are determined experimentally at the oxygen point, the triple point, and the steam point, respectively.

B. From the ice point to the antimony point (630.5°C),

$$R_t = R_0(1 + At + Bt^2) \qquad (6.36)$$

where the symbols are the same, but A and B are to be determined experimentally by the fixed points in this range.

C. From the antimony point to the gold point,

$$E = a + bt + ct^2 \qquad (6.37)$$

where E is the electromotive force at a standard platinum and platinum–rhodium alloy thermocouple, and a, b, and c are determined experimentally from the fixed points in this range.

D. Above the gold point the temperature is defined in terms of the radiative properties of a black body.

TABLE 6.1   Standard Equilibrium Points[a]

Equilibrium State	Temperature			
	Degrees Celsius (°C)	Degrees Kelvin (°K)	Degrees Fahrenheit (°F)	Degrees Rankine (°R)
1. Absolute zero	−273.16	0.0	−459.69	0.0
2. Equilibrium temperature between oxygen and its vapor (oxygen point)	−182.97	90.19	−297.346	162.344
3. Equilibrium between ice, liquid water, and its vapor (triple point)	+0.01	273.17	32.05	491.74
4. Equilibrium between liquid water and its vapor (steam point)	100.00	373.16	212.0	671.69
5. Equilibrium between liquid and solid zinc (zinc point)	419.505	692.665	755.11	1214.80
6. Equilibrium between liquid sulfur and its vapor (sulfur point)	444.6	717.76	832.28	1291.97
7. Equilibrium between solid and liquid silver (silver point)	960.8	1233.96	1761.4	2221.09
8. Equilibrium between solid and liquid gold (gold point)	1063.0	1336.16	1945.4	2405.09
9. Equilibrium between liquid and solid water at 14.6959 psia (ice point)	0	273.16	32.0	491.69

[a] From J. F. Swindells (ed.), *Precision Measurement and Calibration—Temperature*, NBS Special Publication 300, vol. 2, U.S. Department of Commerce, Washington, D.C., 1968.

The definition of the ice point of Table 6.1 cannot be reproduced with accuracies greater than $\pm 0.001°C$. The depth below the surface at the triple point must be considered, since the pressure increases with depth. This depth correction is given by the relationship

$$t = 0.01°C - (0.7 \times 10^{-6} \ °C/mm)h \qquad (6.38)$$

where h is the depth in millimeters. If the water is not pure (more than a single component), the error in the establishment of the fixed standard point is unknown. It is necessary to use distilled (or triple distilled) water for calibration purposes. Many laboratory errors result from using tap water in the establishment of the ice or triple point. This same problem exists at the steam point or at any other equilibrium point. A pure substance is required.

Higher temperature scales and lower temperature scales will be defined when the measurement systems become more accurate. The ideal gas thermometer is limited to the range 0.3–1000°K. The ideal gas scale can be used to fix the equilibrium states to the thermodynamic scale in this range. Outside this range some variation may exist. The reproduction of any of the equilibrium points of Table 6.1 always involves some uncertainties. The design of systems capable of eliminating extraneous inputs is involved. Most laboratories rely on the certification of a measurement system by NBS and use the certified system to calibrate other systems for general laboratory use.

All physical systems are dependent on temperature. Their dimensions, energy content, and vibrational modes vary with temperature. Therefore, any physical object is capable of sensing temperature (the human body is extremely sensitive to temperature). However, the range and observable effects of temperature limits the use of physical objects for the measurement of temperature. Measurement systems are basically divided into three ranges for the determination of temperature: low temperatures, normal temperatures, and high temperatures. These range divisions are arbitrary, and a particular sensor may be capable of measuring in more than one range.

Because of the abundance of temperature sensors, additional limitations are imposed by the considerations of the basic concepts of temperature. Temperature is a manifestation of the molecular activity of atoms. The concept of equilibrium assumes that all molecules have the same activity, However, statistics dictate that there are present both high- and low-energy atoms. Therefore, the measurement of temperature at a geometric point is meaningless. Temperature is a measure of the average molecular activity of atoms in a given space. The size of the space selected is basically dependent on the size of the sensor and the boundary conditions. The size of the sensor should be small if a temperature gradient is to be measured. It can be large if the temperature gradients are negligible.

Temperature is also a measure of the potential for the flow of energy in the form of heat. If the sensor absorbs heat or transmits heat from the system being measured, an extraneous input is introduced. Energy transfer by conduction, convection, radiation, and change in flow characteristics obey the first axiom of measurement. These types of extraneous inputs require serious consideration in the evaluation of temperature-sensing devices. The effects of transient temperature are considered in Chapter 7. However, the basic design of sensors is involved in the averaging effect of the sensor element. The sensor responds to the average temperature of its environment over the overall dimension of the element.

The accuracy of measuring temperature is $\pm 0.0001°C$. This accuracy does depend on the range of the sensor, but the basic limitation is imposed by the accuracy of the fixed points. When this accuracy is compared to the accuracies of measuring time, length, and mass (one part in 10^{-8} or 10^{-9}), the characteristics of temperature measurement are fully realized. Fortunately, this accuracy for the measurement of temperature is acceptable for the present requirements on temperature. The effects of nonequilibrium and the technical knowledge of heat-transfer processes have not been developed to the point where this uncertainty is unacceptable. Hopefully, the measurement sensors for temperature will be more sophisticated when required. Already, the magnetic properties of paramagnetic solids are being used in the measurement of temperatures below $0.3°K$.

MEASUREMENT SENSORS FOR LOW TEMPERATURES

Four sensors will be considered for use in the measurement of temperatures from $0.3°K$ to $273.15°K$: the ideal gas thermometer, the platinum-resistance thermometer, the thermistor, and the thermocouple. These systems represent a spectrum of devices capable of determining temperature in an equilibrium system. There are other sensors available, but the selection is based on present usage.

THE IDEAL GAS THERMOMETER

The ideal gas thermometer operates on the principle that

$$\frac{P_1 V_1}{T_1} = \frac{P_2 V_2}{T_2} \tag{6.39}$$

If a device with constant mass is designed to have a constant volume, the establishment of one temperature and pressure (a fixed standard) and the measurement of a second pressure determines the temperature associated with this pressure. The pressure at the reference temperature can be reduced by evaluating the sensor volume, and the measurement can proceed to the point where continuity of the mass contained is no longer valid. The lower limit of this system is the vapor pressure of helium-3 (He_3). The introduction of He_4 may extend the scale. The sensor is dependent on the volume, which changes with temperature. It is also dependent on the departure of a gas from the ideal representation. At low pressure, the departure from the ideal gas assumption is very small, and the system can function with little error at high-vacuum conditions. The basic requirement to be satisfied is that of maintaining a continuum. The ideal gas thermometer is a basic device for establishing consistence between the fixed-point standard and the thermodynamic temperature scale.

THE RESISTANCE THERMOMETER

The resistance of an electrical conductor is dependent on the temperature of the resistor. For stable resistance, a platinum wire has been employed as the resistance element. The fact that the recommended interpolation techniques for the low and normal temperature ranges incorporate the electrical-resistance thermometer is indicative of the stability and accuracy attainable with this sensor. The construction of resistance thermometers is extremely important. NBS will certify (calibrate and provide the resistance-versus-temperature tables) the resistance thermometer if it meets the proper construction requirements. This certified thermometer can be used as a secondary standard (working standard) over the range of its certification. The resistance of platinum changes very slightly with temperature, and very elaborate resistance-measuring devices must be available. Also, the bulb size is usually relatively large (a cylinder with a diameter of $\frac{1}{4}$ in. and a length of 4 in.) in order for the resistance thermometer to provide an initial resistance that is large enough to produce an adequate signal.

The change in resistance with change in temperature (sensitivity) of the resistance thermometer limits the range from 90.19°K to 717.76°K. This is not an absolute limitation, but three factors influence the limitation to this range: the sensitivity, the current flow in the platinum resistor necessary for the measurement of resistance, and the nonisothermal condition present in the system being measured. These extraneous inputs can be evaluated over the range indicated. However, extension beyond this range must be attempted with caution. The development of precision bridge circuits has enhanced the utility of the platinum-resistance thermometer. A discussion of the modifying and read-out equipment required by this sensor will be given in the following sections of this chapter.

Methods for maintaining the fixed temperature points of Table 6.1 are discussed in reference 2. Analysis of the temperature drop across interfaces is also made. When the maximum accuracy capabilities of the platinum-resistance thermometer are to be developed, these precautions are necessary. (That is, when the temperature is to be measured with accuracy $\pm 0.0001°C$, all extraneous inputs must be evaluated and proper corrections determined.)

The acoustical thermometer and germanium-resistance thermometer have been used to establish a temperature in the range 2–20°K. These devices and their operation are discussed in reference 2. However, they are not discussed here. The germanium-resistance thermometers are reproducible to $\pm 0.001°K$ (reference 2).

The platinum-resistance thermometer can measure temperatures in the range of 1100°C. However, the accuracies in this range are $\pm 1°C$. The cost of these sensors and the fact that their properties do change with extended use at elevated temperatures generally restricts their range to the 630°C range.

The platinum-resistance thermometer requires a very precise measurement of resistance. The resistance of the lead wire used to connect the thermometer to the resistance bridge circuit can cause an error in the determination of resistance. Several methods of correcting for lead-wire resistance are possible. (1) The resistance of the lead wire can be measured. (2) Compensating leads may be added. One disadvantage of simply measuring the lead-wire resistance is that resistance changes with temperature. This error is eliminated by placing an equal length of wire in the adjacent leg of a bridge circuit and providing this wire with the same thermal environment as that of the lead wires connecting the thermometer element (see reference 3, p. 456).

The basic sensor (the platinum wire) is sensitive to mechanical strain, humidity, and thermal gradients. The approximate sensitivity, dR/dT, is in the range of 0.09 $\Omega/°C$. The element is very stable at normal temperatures, and the reproducibility characteristics are the principal reason that it can be used for a secondary standard.

THE THERMISTOR

The thermistor is also a variable-resistance element for sensing temperature. It is a semiconductor device made of a ceramic-type material. The resistance of a thermistor increases with decrease in temperature,

$$R = R_0 \exp\left[\beta\left(\frac{1}{T} - \frac{1}{T_0}\right)\right] \tag{6.40}$$

where R_0 is the resistance at temperature T_0 in degrees Kelvin, R is the resistance at temperature T in degrees Kelvin, and β is a constant. The thermistor resistance has a negative temperature gradient and is exponential in form, while the platinum resistance has a positive temperature gradient and is polynomial in form (see Equation 6.35).

The sensitivity of the thermistor in per cent resistance change per degree Celsius is approximately -4%-Ω/°C. The normal ranges of the constant β are 3400–3900 (reference 3). From Equation 6.40 the sensitivity is

$$\eta = \frac{dR}{dT} = R\left(\frac{-\beta}{T^2}\right) \tag{6.41}$$

and the per cent sensitivity is $\eta/R = -100\beta/T^2$. This device is obviously more sensitive at low temperatures than at high temperatures. The resistances are very high (150,000 Ω) at lower temperatures, while the platinum-resistance element is in the 25-Ω range. Therefore, the difficulty of measuring temperature with a thermistor involves high resistance values.

The thermistor does not have the stability characteristics of platinum wire. This problem has been the major reason for selecting the platinum wire for a working standard. However, after the thermistor has been calibrated, it does provide a very usable temperature-measuring system. It is frequently used in temperature-compensating elements, control systems, and in measuring dynamic signals. It is much smaller in size than the platinum-resistance thermometer and is less expensive. The size and the thermal capacity of ceramic material means that it changes the system being measured to a lesser degree than does the platinum-resistance element.

THE THERMOCOUPLE

Three separate experimental observations led to a formulation of the basic laws governing the operation of thermocouples. In 1821 Seebeck found that a closed circuit of two dissimilar metals (see Figure 6.25) could sustain

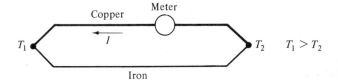

Figure 6.25 Simple Thermocouple Effect

a constant flow of current if the temperatures of the two junctions were different. If T_1 is greater than T_2 of Figure 6.25, the current will flow from copper to iron at T_1 (hot junction), because iron is thermoelectrically positive with respect to copper. The current will flow from iron to copper at the low-temperature junction, T_2. This is called the Seebeck effect.

In 1834 Peltier discovered that the introduction of a current in the direction of Figure 6.25 (where the meter is shown) would cause the junction at T_1 to absorb heat and the junction at T_2 to liberate heat. A reversal of the direction of current flow would cause absorption of energy at T_2 and rejection at T_1. However, the Peltier effect can be essentially neglected when the current flowing is entirely due to the Seebeck effect (or thermal emf generation). This is the Peltier effect.

Thompson discovered that the flow of current through a wire where the temperature varies with position along the wire would cause either the absorption or rejection of heat, depending on the direction of current flow and the thermal gradient. Absorption occurs when current and heat flow are in the same direction (rejection when opposed). The reverse is also true, in that a current flows through a conductor when there is a temperature gradient. This is called the Thompson effect. The Thompson effect actually has less influence on the thermal emf generation than does the Peltier effect. This effect is also generally neglected in the temperature-measuring process.

Three basic laws were formulated, based on the above experimental observations: (1) the law of homogeneous circuits, (2) the law of intermediate metals, and (3) the law of intermediate temperatures. These laws together allow for the measurement of temperature using some form of the thermocouple circuit of Figure 6.25. The law of homogeneous circuits states that there can be no flow of current in a closed circuit composed only of a single homogeneous metal. One test for homogeneous circuits is to measure the current when the circuit is placed in an oven. If a sustained current exists, the circuit must be assumed to be inhomogeneous. The law of intermediate metals states that the current sensed in a circuit composed of any number of different metals is zero if the temperature of the circuit is constant and uniform. This second law of thermocouples allows for the insertion of a meter into the thermocouple circuit, provided the temperature at the two junctions of the circuit and the meter are the same. Under these conditions the total emf generated in the original thermocouple circuit will be identical to that generated in the circuit incorporating the meter. Without the second law of thermocouple circuits it would be very difficult to measure the generated emf. The third law of thermocouples states that the thermal emf generated by any thermocouple circuit composed of homogeneous metals with junctions at any two temperatures $(T_1$ and $T_2)$ is algebraically equal to the thermal emf generated with the same thermocouple when an intermediate temperature is introduced. Figure 6.26 is a statement of the third law. In

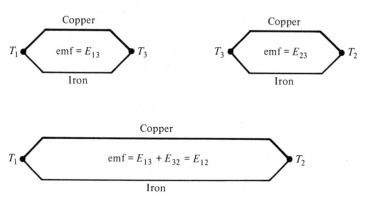

Figure 6.26 Law of Intermediate Temperatures

word form the emf generated by a thermocouple of homogeneous metals with junctions at T_1 and T_3 algebraically added to the emf generated by the same thermocouple with junctions at T_3 and T_2 is equal to the emf generated by this thermocouple with junctions at T_1 and T_2.

The thermocouple is one of the simplest temperature sensors. The requirement that each material in the thermocouple be homogeneous is a serious restraint. However, several manufacturers do supply thermocouple wire that is sufficiently homogeneous. When any alloy is present in thermocouple material, the emf characteristics are changed. Tables of values of emf versus temperature are provided for one reference temperature. This is possible due to the law of intermediate temperature. [Thermocouples and their ranges are copper–constantan (-300–$650°F$), iron–constantan (0–$1500°F$), chromel–constantan (0–$1000°F$), chromel–alumel (100–$200°F$), platinum–10 per cent rhodium (1300–$2700°F$); tables of emf versus temperature are available for others.] If the tables are to be useful in determining temperature, two requirements must be met: the thermocouple material must be homogeneous and the junction must be homogeneous. Tests for homogeneous wire are given in reference 2. The junction can have nonhomogeneous properties if the entire junction is at a uniform temperature (law of intermediate metals).

The temperature measured by a thermocouple is essentially the temperature at the point of electrical contact nearest the measuring system, There are several junction shapes and several methods for joining the junction available. Figure 6.27 gives several typical junction configurations. The

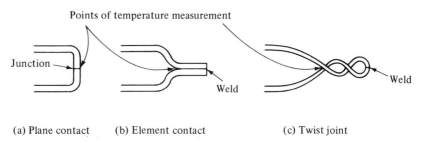

Points of temperature measurement

Junction

Weld

Weld

(a) Plane contact   (b) Element contact   (c) Twist joint

Figure 6.27 Thermocouple Junctions

temperature measurement reflects the temperature at the points illustrated. Four joining methods are mechanical twist, fusion welding, brazing (or soldering), and arc welding. Reference 4 gives a comparison of thermocouple joining techniques. Brazing provides good electrical contact and does not change the properties of the individual thermocouple elements, but it does introduce an intermediate metal. Fusion does introduce some change in the elements because of the elevated temperature required and the elements in the surrounding air. Mercury arc welding under oil can reduce the effects of fusion, but mercury and oil do have contaminants. The mechanical contact is unstable, and the points of contact are limited because of contact resistance. Mechanical contact is probably the least desirable method of joining.

A very brief list of common thermocouple materials in the emf chart are, in decreasing order: chromel, iron, copper, platinum–10 per cent rhodium, platinum, alumel, constantan. Each material is thermoelectrically positive with respect to those below it and negative with respect to those above. (Chromel is positive with respect to all materials listed, and constantan is negative with respect to all materials listed.)

Most tables for thermocouples use the ice point for a reference junction. When temperature is to be measured with accuracies less than $\pm 0.1°C$, this can be an adequate reference point. The circuit for any thermocouple in measurement is shown in Figure 6.28. The reference temperature is com-

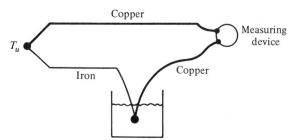

Figure 6.28 Thermocouple Measuring Circuit

monly the ice point, and the measuring device has both lead junctions at the same temperature or is constructed with all wiring of copper for Figure 6.28. The measuring device is usually selected so that no current flows in the thermocouple circuit (to avoid lead-wire resistance and the Peltier and Thompson effects). These voltage-measuring devices are discussed in the following section of this chapter. Equation 6.37 indicates that the general relationship between emf and temperature for a thermocouple is a polynomial. The sensitivity for a thermocouple, $d(\text{emf})/dT$, is positive and approximately 0.024 mV/°F (0.043 mV/°C). Originally, the very small voltage output from thermocouple circuits was a serious limitation to the use of thermocouples as temperature sensors. However, the development of high-gain amplifiers and low-voltage measuring devices has progressed to the point where thermocouple temperature-measuring systems are common.

Two thermocouple circuits commonly used in measurement are the parallel circuit and the series (or thermopile) circuit. These circuits are shown in Figure 6.29. The output voltage for the parallel circuit is equal to $(E_{1-2} + E_{1-3} + E_{1-4})/3$, which is the average emf generated between T_1 and the other three temperatures, T_2, T_3, and T_4. The output voltage for the series circuit is the algebraic sum of the emf's generated between T_1 and T_2 (i.e.,

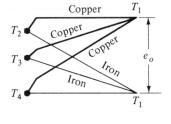

(a) Parallel circuit

(b) Series circuit

Figure 6.29 Typical Thermocouple Circuits

emf $= 3E_{1-2}$). The series circuit is called a thermopile, because it generates a voltage that is equal to the number of thermocouple circuits times the emf generated for one thermocouple between the two junction temperatures.

MEASUREMENT TECHNIQUES FOR THERMOCOUPLES

Thermocouples are among the most commonly used temperature-sensing elements. They are not always completely understood by the user and are frequently major sources of measurement error. Five very important extraneous inputs result from thermocouple resistance: extension leads to the output meter, electrical noise pick-up, stability, accuracy, and ice point. When the output meter is of the type that allows no current flow in the thermocouple circuit, the thermocouple resistance, the Peltier effect, and the Thompson effect are eliminated. If a deflection-type meter is used with current flow, the voltage drop in the thermocouple wire must be considered. The Peltier and Thompson effects will still be very small (negligible) for the thermally induced current.

The extraneous input introduced by the use of extension leads and a measuring device is twofold. First, there is the same resistance problem as that of the thermocouple itself. Both of the resistance problems can be reduced by placing a resistor in series with the circuit when null-type balancing instruments are used. This resistor is large enough that the wire resistance is negligible in comparison. (When two or more circuits in series or parallel are used, the difference in thermocouple length causes error.) The second error resulting from the use of extension leads and measurement circuitry is the introduction of two (or more) junctions of dissimilar metals. (A measurement system with all wire made of copper introduced in the copper side of a thermocouple will not introduce new junctions, provided both copper metals are identical.) One method for reducing or eliminating the extraneous inputs is to maintain equal temperatures at all introduced junctions. Figure 6.30 gives two methods of eliminating the extraneous input. In both cases the circuit in the measurement instrument is to be made of copper wire identical to that of the input leads.

The low resistance of the thermocouple circuit (often less than 1 Ω) and the presence of electrical noise in most mechanical equipment and in the air can cause the thermocouple to function as an antenna. If the noise is high-frequency voltage, the dc sensing instrument can reject the signal. However, when the noise is very low frequency or when it has a dc bias imposed on the signal, the output instrument can sense the extraneous input. For exam-

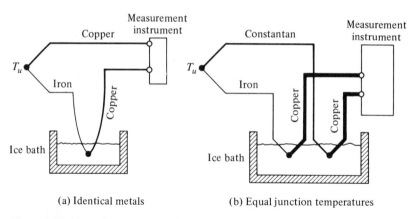

(a) Identical metals   (b) Equal junction temperatures

Figure 6.30 Measuring System Circuitry

ple, tap water heated by a dc electrical system could very easily conduct electrical current and cause an off-ground voltage field. Any type of common ground between the thermocouple circuit and the water could cause a voltage potential to be picked up by the circuit. The sensitivity of the thermocouple is approximately 24 μV/°F. Certainly, stray dc voltages of this magnitude could easily be present in many measuring situations. One method of reducing electrical noise is to provide insulation on all exposed thermocouple wire. Plastic, glass, and enamel coatings are relatively good electrical insulators. A reduction of the thermal response of the thermocouple will result from using insulation. A second method for reducing noise is to be sure that all junctions in the circuit are welded, brazed, or soldered. If the junctions are of the same material, they should also be shielded from the surroundings. One check for noise is to place both reference and measuring junctions at the same temperature (in different containers if possible) and check the output voltage; it should be zero.

Thermocouple range and accuracy depends on the materials and the care taken in calibration. Table 6.2 gives a few of the characteristics of thermocouples. Attempting to obtain accuracies greater than those listed in Table 6.2 could easily result in an introduced extraneous input. The change in stability could also result in an extraneous input. Table 6.2 gives the calibration practices recommended by the NBS. The stability of thermocouple circuits is very good for the temperature ranges suggested. However, single-point calibration should be performed before the thermocouple is used each day. It is always better to check at the beginning of an operational period and at the end to be certain that the system operates properly.

The presence of thermal gradients in ice baths has been known for some

TABLE 6.2 Thermocouple Characteristics[a]

Material	Temperature Range (°C)	Calibration Points	Accuracy at Calibration Points (°C)	Equation Interpolation	Uncertainty in Interpolation Values (°C)
Platinum–platinum 10 per cent rhodium	630.5–1063	Antimony, silver, gold points	0.2	$E = a + bT + cT^2$	0.3
Chromel–alumel	−190–+300	Resistance thermometer	0.1	Tables	0.5
	0–1100	Standard thermocouple	0.5	$E = a + bT$	1.0
	0–760	Standard thermocouple	0.5	$E = a + bT$	1.0
Iron–constantan	−190–+350	Resistance thermometer	0.1	Tabular values	0.5
Copper–constantan	0–300	Resistance thermometer	0.1	$E = aT + bT^2 + cT^3$	0.2
	0–100	Boiling water Ice point	0.05	$E = aT + 0.04T^2$	0.2
	−190–0	Resistance thermometer	0.1	$E = aT + bT^2 + cT^3$	0.2

[a] From J. F. Swindells (ed.), *Precision Measurement and Calibration—Temperature*, NBS Special Publication 300, vol. 2, U.S. Department of Commerce, Washington, D.C., 1968, pp. 254–320.

time. Errors in ice-bath temperature could be 2–3°C. This error is extremely serious if the required measurement accuracy is ± 1°C or less. Reference 2 gives some methods of preparing an ice bath. Ice made from distilled water should be crushed and placed in a well-insulated container (a quality vacuum bottle). Only enough distilled water should be added to fill the voids in the ice. Some ice flotation will occur, but ice should be added and water removed to maintain ice at the bottom of the chamber. The thermocouple should be placed in the bath at a depth where conduction down the wire does not affect the junction temperature. This could be accomplished by several coils of wire around the junction placed in the bath. A more acceptable ice point can be obtained by placing a Pyrex tube 8 or 9 in. into the bath. The tube contains a $\frac{1}{2}$-in. column of mercury in the bottom. A thermocouple junction placed in the mercury will be maintained at the ice-point temperature. For precision baths, an ice bath is surrounded by an ice bath (reference 2).

THERMOCOUPLE TEMPERATURE MEASUREMENT

To measure temperature using a thermocouple, one reference temperature must be known with an accuracy of 10 times the desired temperature accuracy. The emf tables can determine whether the measured temperature is higher or lower than the reference temperature. [This can also be accomplished experimentally by observing the sign ($\pm$) of the output voltage when the measured temperature is higher than the reference temperature; with an ice-point reference junction, the body temperature is higher.] Then, if the sign of the unknown temperature is the same for the same reference junction, the unknown temperature is higher than the reference temperature. If the sign of the unknown temperature is opposite, the unknown temperature is lower. (If the steam point is the reference junction, body temperature is lower, and unknown temperatures with the same voltage sign as that of the body temperature are lower than the reference temperature.) Usually, the unknown temperature is obviously either higher or lower than the reference temperature. This information is necessary when thermocouples are to be used.

EXAMPLE 6.1

If the ice point is used for the reference junction of a copper–constantan thermocouple, and the voltage is -1.517 mV (using the circuit of Figure 6.28 with constantan replacing the iron wire), what is the value of the unknown temperature? The sign of the output voltage is negative, but this sign is determined by the method of connecting the leads to the meter. If the copper lead from the reference junction is connected to the positive terminal of the meter, and the copper lead from the unknown temperature junction is connected to the negative terminal, the negative sign means that the unknown temperature is lower than the reference temperature. If the leads are reversed, $T_u > T_R$. For the latter case, $T_u > T_R$, thermocouple tables give $T_u = 100°F$.

Typical Copper–Constantan Tables (Small Range Selection)

T (°F)	20	40	60	80	100
0	−0.254 mV	+0.171	+0.609	1.057	1.517
100	1.987	2.467	2.958	3.458	3.967

If a second unknown temperature gives $E = -1.567$ mV, what is the unknown temperature and what would be the voltage if the reference temperature was 100°F?

The unknown temperature is higher than the ice-bath temperature. This is based on the first part of this example. It lies between 140°F and 160°F. Linear interpolation between these points gives $T_u = 144.07°F$. The emf $E_{100-144}$ could be found in the tables, but let us use the law of intermediate temperatures:

$$e_{R-100} + e_{100-144.07} = e_{R-144.07}$$

$$e_{100-144.07} = e_{R-144.07} - e_{R-100}$$

$$= -2.567 - (-1.517) = -1.050 \text{ mV}$$

This voltage results when the leads are connected in such a way that the negative sign means the unknown temperature is higher than the reference temperature.

NORMAL TEMPERATURE RANGES

All of the temperature-measuring devices discussed in this section can be used to measure temperature in the range 0.0–200°C. This range will arbitrarily be called the normal range. Five additional temperature-measuring systems will be discussed for the normal range: metal bar, bimetal strips, liquid-in-glass thermometer, pressure-gage thermometer, and quartz thermometer. Discussions of these systems will be abbreviated, although they are very useful systems. Many other temperature sensors are available and in common use. We already know that all physical objects are sensitive to temperature. The expansion or contraction of any material with a change in temperature is the principle of operation of the metal bar, the bimetal strip, and the liquid-in-glass thermometer. The determination of temperature is based only on a length measurement. Figure 6.31 shows these three temperature sensors. For the rod shown, the length is given by $(x - x_0)/x_0 = \alpha(T - T_0)$, where x_0 is the length at temperature T_0, x is the length at any temperature T, and α is the thermal coefficient of linear expansion.

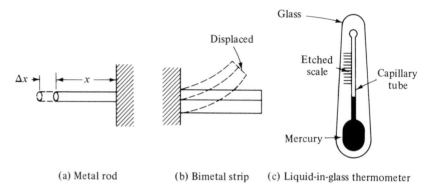

(a) Metal rod   (b) Bimetal strip   (c) Liquid-in-glass thermometer

Figure 6.31 Temperature Sensors

EXAMPLE 6.2

At 68°F a stainless steel rod is 5 in. long. The thermal coefficient of expansion is approximately $9.1 \times 10^{-5}/°F$. What is the length of the rod at 69°F? 1000°F?

At $T = 69°F$,

$$\frac{x - x_0}{x_0} = \frac{x - 5}{5} = 9.5 \times 10^{-6} \qquad \text{(reference 1)}$$

$$x = 5 + 4.75 \times 10^{-5} = 5.0000475 \text{ in.}$$

At $T = 1000°F$,

$$x = 5 + 5 \frac{9.5 \times 10^{-6}}{°F} \, 932°F \text{ in.} = 5.04425 \text{ in.}$$

To measure a temperature change of 1°F requires a length-measuring device capable of measuring $\pm 10^{-7}$ in. The precision length-measuring system (possibly interferometer) required by the metal-rod sensor makes the system too expensive. The range of this device is (approximately) -300–2000°F. However, accuracy is a problem.

A bimetal strip is formed by bonding dissimilar metals having different thermal coefficients of expansion. The method of bonding should ensure that no relative motion between the metals occurs for the range of temperatures to be measured. (Brazing is one possible method of joining.) The strip will be flat at some initial temperature, which is usually the bonding temperature unless the bonding takes place with prestressed elements. When the temperature of the bimetal element is different from the initial flat-condition temperature, the metal with the larger coefficient of expansion will tend to be longer for higher temperatures and shorter for lower temperatures. The only way one side of the strip can have a longer length than the other side is for the metal strip to curve into an arc of a circle. In Figure 6.31 the dashed

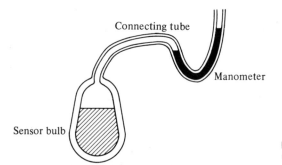

Connecting tube

Manometer

Sensor bulb

Figure 6.32 Pressure-Gage Thermometer, Liquid-Vapor

lines represent the displaced position for the bimetal strip. The range of the bimetal strip is approximately -100 to $+1000$°F. This device is particularly useful in control systems where the mechanical motion can be used to close an electrical circuit. The thermostatic control devices in many buildings are of this type. Bimetal strips are structurally strong, relatively insensitive to the environment, and very reliable. The devices are not used actually to measure temperature. They sense the change in temperature, and these changes are used to actuate a system. To measure actual temperature with this system would require the measurement of the radius of curvature of the displaced bimetal strip. The difficulties involved in this measurement essentially restrict the bimetal strip to use in situations where the actual value of temperature is not desired.

The liquid-in-glass thermometer is the most common temperature-measuring device. In this case familiarity leads to unjustified faith in the instrument. The principle of operation is the difference in the thermal coefficient of volumetric expansion of two dissimilar materials. Typical combinations are mercury in glass, alcohol in glass, and mercury and nitrogen in glass. If Pyrex glass is used, its thermal coefficient of volumetric expansion is very small compared to the liquids used to fill the bulb and capillary tube. The liquid-in-glass thermometer measures the average temperature adjacent to the thermometer. Total-immersion and partial-immersion thermometers are available. Nonlinarities are present when any portion of the thermometer is exposed to temperatures other than that of the rest of the thermometer. Corrections for immersion errors are available for each type of liquid-in-glass thermometer. One problem inherent in the mercury thermometer is the requirement for an isothermal environment over the whole thermometer. The bulb is the most sensitive portion of the thermometer, but the heat transfer by conduction down the glass–mercury system can alter the temperature at the bulb.

The range of liquid-in-glass thermometers is approximately -50 to $+800°F$. The accuracy obtainable in an isothermal environment is $\pm0.5°F$ ($0.278°C$). One advantage of the mercury-in-glass thermometer is that the sensor and output device come in one package. The etched scale on the thermometer gives the output temperature sensed. The simplicity of operation tends to make one overlook the fact that extraneous inputs are also present with this system. This type of system should not be used to measure a temperature gradient or to measure the surface temperature of a metal plate. The total heat-transfer problem must be solved in these cases, and this solution can include many error sources.

The glass thermometer is also very sensitive to external pressure, since the increase in pressure tends to compress the bulb and produce an effect identical to an increase in temperature. The vapor pressure of the mercury and any inert gas above the mercury are error sources. There is eventually a change in the volume of the glass with thermal cycling. If the effects are permanent, the thermometer must be calibrated for this new condition. One of the main disadvantages of the glass thermometer is that it is fragile. However, most sensors are relatively fragile. When an accuracy of $\pm1-3°C$ is acceptable, the glass thermometer is a very simple and useful sensor. When more refined temperatures are required, extraneous inputs must be considered seriously.

The pressure-gage thermometer is composed of a sensing bulb, a connecting tube, and a pressure gage. Figure 6.32 shows one possible arrangement for a pressure-gage thermometer. The connecting tube will always be at some temperature other than the bulb. This is an extraneous input for total liquid and total gas filled systems. If the sensing bulb and connecting tube contain liquid throughout, the change in pressure measured is essentially the change in the volume of the bulb with temperature. (The change in liquid density is small in comparison, but it is an extraneous input.) When a gas is contained throughout, the pressure change is primarily a result of the change in the pressure with change in temperature. However, the density of the gas and bulb volume changes are also present, introducing other inputs.

A system filled only by a liquid and its vapor obeys the conditions for establishing the thermodynamic state of a simple compressed substance. When two phases are present, temperature and pressure are not independent thermodynamic properties. Therefore, knowing the fluid and the pressure determines the temperature. This type of system reduces the problems of extraneous inputs, but condensation of the vapor in the connecting tube can change the system.

The sensitivity of a water–vapor system varies with temperature, but at 14.7 psia the sensitivity is approximately 0.3 psia/°F. The range of this type of device is -300 to $+700$°F.

The quartz thermometer uses the principle that the resonant frequency of a quartz crystal varies with temperature. When the crystal is cut with the proper angle, the frequency varies linearly with temperature. In reference 4 the resolution is said to be ± 0.0001°C in the range -40 to $+230$°C, and the resolution time is 10 s. The basic sensor still has the heat-transfer problem to consider, but it seems that the resolution of this system is an answer to the need for a precision temperature-measuring system. This resolution is almost to the point where the reproducibility of the fixed points of the temperature scale becomes the limit on accuracy.

HIGH-TEMPERATURE RANGES

The high-temperature range is above 200°C. Every sensor previously discussed can be used in this temperature range, at least at the lower values. The thermocouple and the platinum-resistance thermometers are accepted interpolation methods between fixed points in this range. Therefore, the sensors to be introduced in this section are those for very high temperature. The phase or property change of materials, optical pyrometer, total-radiation pyrometer, and spectral analysis methods will be discussed for completeness. Several different types of material are used for fixed-point (or fixed-interval) temperature sensing. The properties of several types of material provide observable melting points, change of color points, and change of shape that can define temperature within a few degrees. The Tempil Corporation of New York supplies crayon, lacquer, and pellets that appear chalky up to the melting point and then appear shiny at the melting temperature. These sensors range from 110 to 2000°F in steps of approximately 50°F (reference 3). The Curtiss–Wright Corporation of Clifton, New Jersey, distributes similar indicators, which produce particular color changes at a distinct temperature. And the ceramic industry has long used seger cones (which melt with cone tip curl at given temperatures) to indicate the temperature at which a particular produce is to be fired. The range of these cones is 1100–3600°F in steps of approximately 50°F (reference 3). Although these devices do not provide continuous temperature scales, they can provide 10–50°F ranges to be used in maintaining a furnace temperature within prescribed limits.

The total-radiation pyrometers and optical pyrometers use the radiant-energy characteristics of materials at high temperatures. The basic sensor for the total-radiation pyrometer is a thermopile. Figure 6.33 is a schematic of the elements of this system and those of the optical pyrometer. The total-radiation pyrometer receives a focussed light wave from the source and uses this energy to heat a collector plate where the hot junction of a series thermocouple circuit is located. The cold plate is thermally insulated from the receiving plate, and the resultant output voltage is a result of the temperature difference between these plates. The receiver-plate temperature is primarily dependent on the intensity of the energy received, which is in turn dependent on the geometric viewing angle and the distance of the lens from the source. Errors include convection-heat transfer from the hot plate, conduction to the cold plate through the thermocouple wire, and heating of the cold plate. For a fixed viewing angle and distance from the source, the total radiation is dependent only on the energy emitted by the source. This is a function of the emissivity of the source and the temperature of the source. Additional optical focusing equipment and internal heaters may be employed in the more refined pyrometers to increase the energy absorption and to reduce the heat loss. The pyrometer is usually used at temperatures above 550°C.

If the energy from the above system is focused to a photoelectric cell, the only change in the system is that the intensity of the incoming wave is measured instead of the heating capabilities. However, this addition provides accuracies of ±0.06°C at 1063°C, ±0.12°C at 1256°C, ±0.7°C at 1330°C, and ±2°C at 3525°C (reference 2).

The optical pyrometer uses the color sensitivity of the human eye to com-

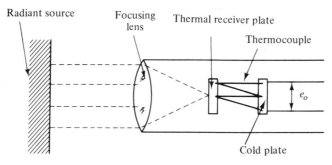

(a) Total-radiation pyrometer

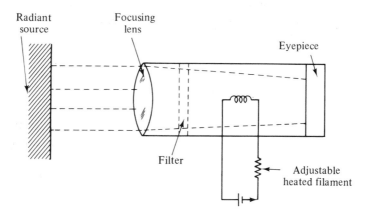

(b) Optical pyrometer

Figure 6.33 Pyrometers for Temperature Measurement

pare the color of an adjustable heated filament to the color of a source background. Filters are provided to reduce the intensity of a high-energy source and to protect the eye. Accuracies of $\pm 0.15°C$ are possible for experienced operators, and standard source signals are available for calibration. The changing characteristics of the heater filament and filament cooling introduce some error. Reference 2 has a relatively complete description of this type of sensor, along with NBS recommendations for its use.

The spectral lines produced by particular molecules at a particular energy level is another source for measuring temperature. A high-energy gas will produce several different spectral lines. Each line essentially represents a given level energy in that gas. At the point at which a new spectral line appears, representing a higher energy level than any previously observed, the temperature of the gas in equilibrium can be evaluated. The thickness of the flame, the continuous spectrum, and the light absorption are extraneous inputs. Again sources and measurement systems are given in reference 2. This reference also includes a very comprehensive bibliography on the field of temperature measurement. If any doubt concerning operating principles, accuracy, calibration, and sources remain, reference 2 should be consulted.

6.6
Measuring Systems for Electrical Quantities

The electrical dimensions current, voltage, and resistance can also be derived from the previously defined standards and Ohm's law. One method of accomplishing this is to define current in terms of the mass of metal plated on a plate in an electroplating medium for a given time. (This, of course, assumes a constant flow rate of electrons from the current source.) Then the requirement that mechanical work be equal to the electrical work, current squared times resistance, defines this in terms of a force applied through a distance in a unit time. Then, electrical work is also equal to current times voltage from Ohm's law, and voltage is defined. Therefore, we can define and measure all electrical quantities directly in terms of the basic standards already defined. The difficulty in this type of approach is that it does not provide the accuracy and precision desired for electrical quantities.

In the analysis of measurement systems for the standard quantities already defined, the sensor, modifying elements, and read-out devices were essentially considered to be one unit. The cost and dynamic response were not of primary importance, because the systems were static and the cost of providing a standard is generally an absolute necessity. The extraneous inputs were pointed out, but the methods of evaluating them were relatively simple for length, mass, time, and force. The section on temperature measurement was considerably longer, because the evaluation of extraneous inputs was much more involved. Electrical quantities are perhaps even more involved, since the standard sources must, in fact, change with usage. Also, these quantities must be compared with corresponding quantities in all other laboratories. This comparative job is much too elaborate to be undertaken by individual laboratories. Therefore, each nation maintains a set of electrical standards and a set of working standards. The values assigned for voltage, current, and resistance are defined in terms of basic standards by the experimental efforts of the national laboratory. The absolute accuracies of these assigned values must be less than the accuracies inherent in the basic standards. The national laboratory (NBS) then provides individual laboratories with certified, calibrated standards. These standards are completely reliable when used according to calibration instructions, but they should be returned to NBS for check calibration, usually every 2 years. (Each standard contains instructions on calibration requirements.)

For the first time, we have encountered an absolute standard that is defined in terms of extremely complex experimental operations with rigorous requirements on the control of extraneous inputs. References 5, 6, and 7 contain many of the considerations that must be made in establishing these standards. They also contain a bibliography of hundreds of articles that must be read to realize the extent of the precautions and techniques that must be observed to obtain electrical standards. The engineer can easily become overwhelmed by this mass of work and happily accept the standards provided by NBS. The only thing wrong with this arrangement is that the lack of complete understanding of a standard frequently leads one to completely block out this massive body of knowledge. Then one is likely to pick up a voltmeter, connect the input leads, read the pointer scale, and write the value of "absolute" voltage. This is the most common and the greatest mistake made in experimental programs today. We will accept the standards provided by NBS, but we must be familiar with voltage-, current-, and resistance-measuring devices, at least to the point where the value read from such devices can be used to obtain the true value of the electrical quantity plus or minus some accuracy.

The rest of this chapter will be devoted to the analysis of measurement systems for voltage and resistance, with some comments on current measurement. Ohm's law provides for the determination of one electrical quantity if the other two are measured. Resistance- and voltage-measuring devices can be very accurate, stable, and inexpensive when compared to current-measuring systems. However, if voltage is measured within known accuracy limits, current is theoretically measured by the measurement of resistance within its accuracy limits. This fact allows us to be concerned principally with the measurement of only two electrical quantities, voltage and resistance.

6.7
Voltage-Measuring Systems

The measurement of voltage is considered under two major classifications: direct voltage (dc), and alternating voltage (ac). A basic sensor for dc voltage (current) is the D'Arsonval meter of Figure 3.5. The measurement of ac voltage can be accomplished using a modifying circuit and this sensor. The measurement systems for voltage will be considered first for dc voltage, then for ac voltage.

The D'Arsonval-meter movement is used in most precision dc voltage-measuring devices. Two conditions can exist in the operation of this meter: zero current flow (null detector) and fixed current flow (deflection detector). The operation of this device with variable current flow is not considered here, since other systems are available for this type of operation. The sensitivity of the D'Arsonval meter is limited by three parameters: the spring constant of the suspension wires, the number of coils wound on the core, and the strength of the magnetic field. The sensitivity is, therefore, limited by the physical characteristics of the detector. The spring constant must be strong enough to support the coils and the pointer mechanism. The magnetic field is limited by the magnetic properties of the permanent magnet, and the number of coils is limited by the coil diameter and weight. It is possible to amplify the input signal using various types of dc amplifying circuits, and it is possible to amplify the pointer output by using a light source and a mirror attached to the coil. Figure 6.34 shows these modifications. It should be used in a dark room only.

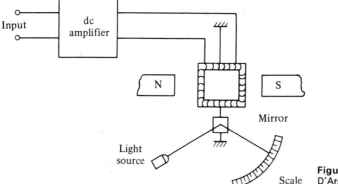

Figure 6.34 Modifications to the D'Arsonval Meter

The D'Arsonval meter (also called a galvanometer) is the basic output device for several different types of electrical measuring systems: voltmeter, ohmmeter, current meter, vacuum-tube voltmeter, and voltage potentiometer (the ohmmeter will be discussed later). The distinction between these meters is the intermediate electrical circuitry. The basic circuits of the voltmeter and current meter are given in Figure 6.35. The basic D'Arsonval meter is constructed with a fixed resistance of the electrical coils, and a fixed current, I_m, will cause full-scale deflection. One precaution must be taken when using this device: the resistors must be selected so that a current greater than I_m never flows through the coils. If this current is allowed to

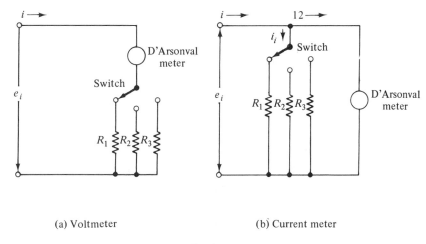

(a) Voltmeter   (b) Current meter

Figure 6.35 Voltmeter-Current Meter Circuits

flow, either the spring support may be broken (or permanently strained) or the coils may overheat and burn out. A voltmeter similar to that of Figure 6.35 should always be set with the largest resistor in series with the meter (select the voltage scale with the largest value available). For an unknown input, e_i, the current flow is

$$i = \frac{e_i}{R_1 + R_m}$$

where R_m is the fixed meter resistance. If the maximum current allowable in the meter is 0.001 A and the meter resistance is 2 Ω, a 1-V input signal would require that R_1 be greater than or equal to

$$R_1 \geq -R_m + \frac{e_i}{I_m} = -z\ \Omega + \frac{1}{0.001}\ \Omega = 998\ \Omega$$

If $R_1 = 900\ \Omega$, the current in the meter is

$$i = \frac{1\ \text{V}}{902\ \Omega} \simeq 0.00111\ \text{A}$$

This would cause a deflection greater than full scale.

The current meter of Figure 6.35 should select the smallest resistor (scale for maximum current reading) to allow a shunt path for current flow. For the same meter properties discussed above, the volt input, R_1, can be determined from Ohm's law for the meter:

$$e_i = I_m R_m = i_1 R_1$$

Therefore, $e_i = (0.001 \text{ A})(2 \text{ }\Omega) = 0.002$ V. The total amount of current, i, that can flow through this circuit is dependent on the shunt resistance, R_1. If R_1 equals 0.01 Ω,

$$i_1 = \frac{0.01 \text{ }\Omega}{0.002 \text{ V}} = 0.2 \text{ A}$$

and $i = i_1 + I_m = 0.201$ A. Usually, ammeters for higher current flow have a larger meter resistance and a larger current-carrying capacity. This is primarily because the selection of smaller shunt resistors is limited by electrical contact resistance.

In both of the above meters, the mechanical contact in the switches causes errors in resistance, especially when dirt, oil, and water accumulate. Operation in the vicinity of large magnetic fields or magnetic materials can cause errors in the meter. After several operations the spring may be subject to yielding, and zero adjust should be made. Electrical noise is present in all laboratories, and measurements of millivolts, microvolts, milliamperes and microamperes can be adversely affected by this noise. Grounding and shielding must be carefully provided for these ranges.

The range and sensitivity of both systems of Figure 6.35 are determined by the value of the resistor selected. The selection of a megohm resistor (10^6 Ω) for the voltmeter would allow the input voltage to be 1000 V, since $i = 1000$ V/1,000,002 $\Omega < 0.001$ A. However, the sensitivity is

$$\frac{di}{de_i} = \frac{d}{de_i} \frac{e_i}{R_1 + R_m} = \frac{1}{1,000,002}$$

This is too small to give an accurate voltage measurement, since change in output is approximately one-millionth the change in the input signal. The maximum sensitivity occurs when the resistance, R_1, is zero since R_m is fixed:

$$\eta = \frac{1}{R_1 + R_m}$$

$$\frac{d\eta}{dR_1} = -\frac{1}{(R_1 + R_m)^2} = 0$$

The optimum sensitivity is determined by this equation when $R_1 = \infty$. This is obviously minimum sensitivity. Therefore, the maximum occurs at the boundary $R_1 = 0$. (The resistance of the lead wires is actually never zero). For the current meter,

$$e_i = i \frac{R_1 R_m}{R_1 + R_m} = i_1 R_1 = i_2 R_m$$

$$\eta = \frac{di_2}{di} = \frac{d}{d_i} \frac{iR_1}{R_1 + R_m} = \frac{R_1}{R_1 + R_m}$$

and the maximum sensitivity based on the selection of R_1 occurs when $R_1 = \infty$:

$$\frac{d\eta}{dR_1} = \frac{1}{R_1 + R_m} - \frac{R_1}{(R_1 + R_m)^2} = \frac{R_m}{(R_1 + R_m)^2} = 0$$

This has a solution when $R_1 = \infty$, and this is the maximum sensitivity condition. (The sensitivity is zero on the boundary with $R_1 = 0$.)

All measurement systems using only resistance-type modifying circuits change the system being measured, because relatively large amounts of current are required to provide an output. For voltmeters a system of designating the current flow required is the rating in ohms per volt. A meter requiring 0.001 A (1 mA) has a rating of 1000 Ω/V. More expensive galvanometers may be built which require only 50 μA (SONA) with a rating of 20,000 Ω/V. The higher the ohms-per-volt rating, the lower will be the current drain on the circuit being measured.

The vacuum-tube voltmeter has a D'Arsonval meter for an output device, but the intermediate modifying circuits are electronic amplifiers. The variety of linear electronic amplifier circuits is so wide that only the very basic concepts are presented here. Actually, the voltage being measured is impressed on the grid-biasing circuit of an electron tube. Figure 6.36 shows one method for accepting the input voltage to be measured. There is essentially no current flow in the grid circuit, because it is negative with respect to the electron source (cathode). Electrons flow from the heated cathode to the anode or plate and then through the external load circuit. The number of electrons flowing is regulated by the intermediate grid circuit. Heating the cathode releases a cloud of electrons in the space above the cathode. The positive voltage of the plate attractes the electrons, while the negative voltage of the grid tends to repel them back to the cathode. The grid becomes less negative with respect to the cathode when a positive input signal occurs. This change in grid bias allows more electrons to escape the grid field and be attracted by the plate. The increased current flow causes a larger voltage drop across the load resistor. The output voltage may be used to drive another biasing circuit or to drive the D'Arsonval-meter movement. In any

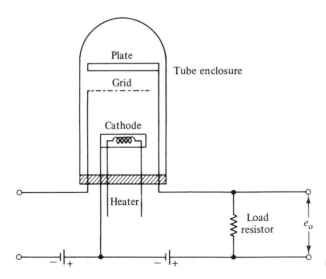

Figure 6.36 Electron Tube Circuit

case the input current is essentially zero and a typical rating for the vacuum-tube voltmeter might be 200,000 Ω/V or more.

The increased resistance and the amplification of the signal are two major advantages of the vacuum-tube voltmeter. However, the problem of electrical noise is much more severe. Also, the current and voltage surges in the electrical power supply can cause serious extraneous inputs. (When a motor starts, the voltage drop in all outlet circuits is apparent by the dimming of the lights.) Electron tubes can operate in ranges that are not linear, resulting in a nonlinear output on the scale. With careful use, an understanding of the limitations, and frequent calibration the advantages of the vacuum-tube voltmeter far outweigh the disadvantages. The cost of these meters is much more than for the simple resistance-type meters.

The voltage-balancing potentiometer is another D'Arsonval-meter application. The act of balancing the voltage to be measured against a known variable voltage requires that no detectable current flows in the circuit being measured. The ohms-per-volt rating can be as high as the sensitivity of the galvanometer to current flow (possibly 10^6 Ω/V). The circuit of Figure 6.37 shows this system. The operation first involves calibration with a standard cell with switch S_3 in position 1. Switch S_1 is closed and the calibration slider is adjusted until the meter reads zero. Then switch S_2 is closed for a more sensitive adjustment of the calibration slider. The slide wire is now calibrated. With switch S_3 in position 2, the input voltage can be balanced with the slide wire over the range of the slide wire. Switches S_1 and S_2 are spring loaded to be open. Switch S_1 has low sensitivity and switch S_2 has

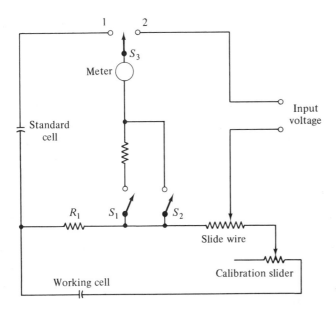

Figure 6.37 Voltage Balancing Potentiometer

high sensitivity. The slide wire is always adjusted first with S_1 closed; then fine adjustment is made with S_2 closed. This particular device is very accurate for measuring the output voltage of a thermocouple, because the zero current flow in the thermocouple circuit is essential. Additional fixed-resistor selections can be introduced in the galvanometer circuit to change the range and sensitivity of the device.

When the standard cell is certified by the NBS, this system can serve as a secondary standard for the calibration of voltage-measuring systems. Extraneous inputs are temperature, mechanical contact resistance, and electrical noise. The normal maximum voltage for the system is 1 V. Protection against higher voltages and ac voltages should be provided by first checking the unknown voltage input with a vacuum-tube voltmeter. Then switch S_1 should be depressed only long enough to ascertain whether or not the galvanometer stays on scale. Once the reading is brought on scale by slide-wire adjustment, the switch may be closed for final adjustment to zero and S_2 can be closed for fine adjustment. This ensures against permanent damage to the galvanometer.

Three other meter principles are used in the measurement of voltage: electrostatic movement, digital converters, and electron deflection. The electrostatic forces developed when two parallel partial discs (electrically insulated) are subjected to a (dc or ac) voltage field will tend to change the relative position of the disc. If one disc is fixed and the other is free to move but spring loaded, the relative torque stored in the spring is proportional to the voltage field impressed. For voltages greater than 100 V, the forces can be accurately determined. This device is one meter which truly measures voltage. Others measure current and use a series resistor. Torque developed is proportional to voltage squared. The digital converter is used to ask logical comparative questions about an input voltage. If an input voltage of 251.75 V is introduced, the converter compares this voltage with internal voltages (100, 200, 300, 400, . . .) until the difference between the input signal and the internal voltage is positive (i.e., 251.75–400 no, 251.75–300 no, 251.75–200 yes). Then this difference (51.75) is introduced in the next lower decade voltages (51.75–100 no, 51.75–90 no, 51.75–50 yes) until the final significant voltage of 0.05 V is determined. Other forms of digital conversion are possible, but all are calibrated using standard voltages. Extraneous inputs are temperature, internal voltage control, humidity, and electron-tube operation.

One of the most important methods of measuring voltage is by electron deflection in a potential field. This method of operation is discussed in detail because it is the single most important method for measuring ac voltages. Therefore, the following discussion will also include the measurement of ac voltage. If an electron is free to move without striking other molecules (in vacuum), it will move in a straight line unless acted upon by some external force (a voltage field); the cathode-ray oscilloscope operation is based on electron motion in a vacuum. Figure 6.38 shows the basic elements of a cathode-ray tube. If no voltage difference exists between any of the deflection plates, the beam of electrons will strike the geometric center of the oscilloscope screen. The cathode produces the free electrons, the

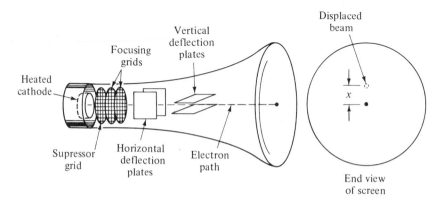

Figure 6.38 Cathode Ray Tube

supressor grid allows control of the number of electrons in the beam, the focusing grids collimate and accelerate the beam of electrons, and the deflection plates provide voltage fields in the path of the electrons. If the top plate in the vertical deflection plate is positive with respect to the lower plate, the beam of electrons will be displaced upward on Figure 6.38. The screen has a fluorescent substance on it to allow the beam contact point to show a spot of light. The displacement of the beam from the geometric center is proportional to the voltage impressed on the deflection plates.

The cathode-ray oscilloscope contains the cathode-ray tube plus electronic circuitry to heat the cathode, control the intensity, control the focus, provide an internal variable saw-tooth wave for the horizontal deflection plates, and provide ac and dc amplifiers for the signal to the vertical deflection plates (also for the horizontal deflection plates). The basic internal circuitry for the oscilloscope is given in Figure 6.39. Circuitry for energizing the cathode-ray tube is not included, although it is part of the internal

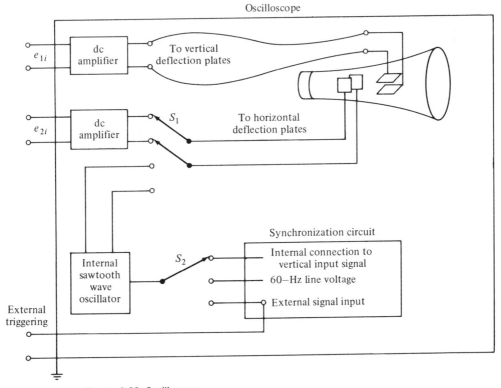

Figure 6.39 Oscilloscope

circuitry. If the voltage input to the deflection plates is impressed directly to the plates with no intermediate amplifiers, this device is also a true voltage-measuring system, since no current is required to flow. However, the use of internal amplifiers gives the same operation characteristics to the oscilloscope as that of the vacuum-tube voltmeter (approximately 10 MΩ resistance). For measuring a dc signal, the sawtooth wave is not required for the horizontal deflection plates. All systems discussed so far have been for the measurement of static dc signals. A brief summary will be given for these systems before changing dc measurement and ac measurement is presented.

Cadmium-saturated cells (voltage $= 1.018636$ V) may be purchased as standard voltage sources. The current flow from this source must be less than 100 μA, the temperature should be maintained at approximately $20°C$, and the emf should be checked yearly. Voltage reference sources based on this cell can be used to calibrate the voltage range from 10 mV full scale (± 0.1 μV) to 100 V full scale. These sources can be used to calibrate the D'Arsonval-meter movement with any associated circuitry discussed. They can also be used for electrostatic-movement meters, digital voltmeters, and oscilloscopes. The accuracy of voltage measurement required dictates the type of measurement system required and the calibration necessary. No meter automatically measures voltage regardless of its precision or cost. It must be calibrated before measurement is meaningful.

The measurement of a changing (with time) dc voltage and an ac voltage can be accomplished using the oscilloscope with an internal sawtooth wave oscillator. The sawtooth wave is shown in Figure 6.40, along with

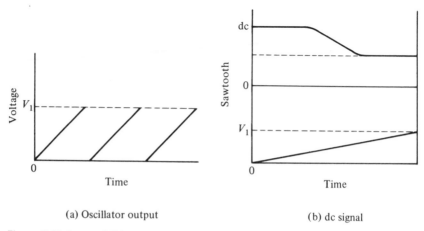

(a) Oscillator output　　　　　　　　　　(b) dc signal

Figure 6.40 Sawtooth Wave

the proper selection of slope to present a transient dc signal on the face of the oscilloscope. Voltage V_1 represents the voltage required to move the beam of electrons from the extreme left of the oscilloscope screen (of Figure 6.38) to the extreme right. The connection of the sawtooth wave to the horizontal deflection plates makes the horizontal scale a time scale. The voltage output is $V = at$, where a is a constant that can be selected; in Figure 6.40(b) the constant was selected so that the total time of the dc signal was less than the time required for 1 cycle of the sawtooth wave. The dc plot of this figure will appear on the face of an oscilloscope if the dc voltage is applied to the vertical input amplifiers, the correct sweep rate is selected (the slope of the sawtooth wave is the constant a and determines the sweep rate), and both sawtooth wave and dc signal start at the same point in time. This last requirement is met by synchronizing the signals.

Synchronization of signals on the oscilloscope may be accomplished in three ways (see Figure 6.39). When single-sweep operation of the oscilloscope is selected, only 1 cycle of the sawtooth wave is allowed on the horizontal deflection plates. The time when this signal is allowed to be impressed on the plates is selected by one of the three modes available for synchronization. The electrical signal which controls the start of the sawtooth wave is called the triggering signal. This operation is identical to the triggering level and slope for the electronic counter (see Figure 6.24). If the signal to the vertical deflection plates is of the proper form, it can be selected to trigger the sweep. For this mode of operation, switch S_2 of Figure 6.39 is turned to the internal signal. When the switch selects a 60-Hz line, the sweep starts 60 times per second (each time the line voltage rises above the selected triggering level for positive slope). Finally, any external signal can be introduced to trigger the sweep.

When the sawtooth wave is the input to the horizontal deflection plates, the trace on the oscilloscope screen is a plot of vertical input voltage versus time. This allows the oscilloscope to measure transient dc signals and ac signals. This same type of plot may also be obtained from recording oscillographs and magnetic tape recorders. Recording oscillographs replace the pointer of the D'Arsonval meter with an ink pen (or thermal device or light source–mirror) and replace the scale with graph paper (or thermal-sensitive paper or light-sensitive paper) which moves under the pen with constant velocity. The dynamic characteristics of this system are given by Equation 3.1. The magnetic tape operates in the same manner, except that the pen is the magnetic head and the paper is magnetic tape. The time base of the oscilloscope, the oscillograph, and the magnetic tape recorder must be calibrated with station WWV or with some laboratory standard.

The dynamic response of the oscillograph with pen or thermal writer is fairly good up to 100 Hz. The light oscillograph may extend up to 500 Hz. Magnetic tapes have responses up to 50,000 Hz, with tape speed being a critical problem. The oscilloscope is limited in response theoretically by the speed of light, but practically the beam intensity and screen persistence limits response to approximately 100,000 Hz. All of these devices must be calibrated for dynamic response by the methods outlined in Appendix III.

The oscilloscope may also be used to plot the input to the vertical plates versus the input to the horizontal plates. When switch S_1 in Figure 6.39

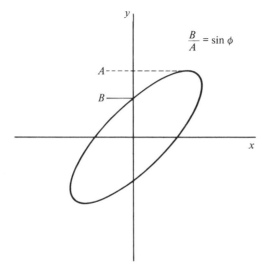

Figure 6.41 Phase Angle Determination

is connected to the input signal, e_{2i}, and x–y plot results. This is the method of operation for obtaining Lissajous patterns. The phase angle between the two signals can be determined from this plot. Figure 6.41 shows a Lissajous pattern for two sine wave signals having the same frequency. When the beam is positioned so that $x = y = 0$ for zero input to both sets of plates, the phase angle between the two signals, ϕ, is given by sine $\phi = B/A$.

To complete the discussion of voltage measurement, the fundamental concepts of ac voltage will be considered. The measurement of high voltages (1000 V and above) are not discussed (see reference 6). Usually, the ac voltage to be measured has a steady-state wave form. Most common voltages (line voltage in our homes) are sine waves. The basic link between ac and dc voltages is the dynamometer movement. This system replaces the permanent magnet of the D'Arsonval meter with a pair of electrical field coils. When current flows in these coils, a magnetic field is developed. If the same current flows through the coils on the core of the D'Arsonval meter, the torque developed is proportional to the current squared. If a resistance is added in series with this system, the movement is proportional to the voltage squared. Over any time interval, the mean value of the reading represents the mean value of the voltage squared. The read-out scale is a square-root scale for this device, and the value read represents the root-mean-square (RMS) voltage. Since the device functions properly for both ac and dc current, it is possible to evaluate ac voltage (RMS) in terms of dc voltage. The dynamometer is a high-accuracy, low-frequency, true RMS meter. If the input signal is a sine wave, $E = A \sin \omega t$, the RMS voltage is obtained from the equation

$$
\begin{aligned}
V(\text{RMS}) &= \left[\frac{\omega}{2\pi} \int_0^{2\pi/\omega} (A \sin \omega t)^2 \, dt \right]^{1/2} \\
&= \left[\left(\frac{A^2 \omega}{2\pi} \right)^{1/2} \left[\left(\frac{t}{2} \right) (\sin 2\omega t) \Big|_0^{2\pi/\omega} \right]^{1/2} \right] \\
&= \left(\frac{A^2 \omega}{2\pi} \right)^{1/2} \left(\frac{\pi}{\omega} \right)^{1/2} = \left(\frac{A^2}{2} \right)^{1/2} = 0.707A
\end{aligned}
$$

The true RMS voltage for various wave forms is shown in Figure 6.42. For the triangular wave, $\frac{1}{4}$ cycle is used to simplify the mathematics. From

$E = A \sin \omega t$
$E(\text{RMS}) = 0.707\,A$

$$
E = \begin{cases} A & 0 \leqslant t 5 \dfrac{\pi}{\omega} \\[2mm] -A & \dfrac{\pi}{\omega} \leqslant t 5 \dfrac{2\pi}{\omega} \end{cases}
$$

$E(\text{RMS}) = A$

$E = \dfrac{2A\omega}{\pi}\, t \quad 0 \leqslant t \leqslant \dfrac{\pi}{2\omega}$

$E(\text{RMS}) = 0.5774\,A$

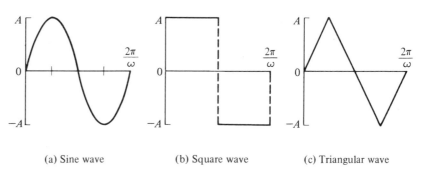

(a) Sine wave   (b) Square wave   (c) Triangular wave

Figure 6.42 Waveforms and their RMS Value

symmetry, the RMS value obtained for $\frac{1}{4}$ cycle is the same as that for $\frac{1}{2}$ cycle, 1 cycle, 2 cycles, and so on. For the triangular wave,

$$V(\text{RMS}) = \left\{ \frac{2\omega}{\pi} \int_0^{\pi/2\omega} \left[\left(\frac{2A\omega}{\pi} \right)^2 t^2 \right] dt \right\}^{1/2}$$

$$= \left[\left(\frac{2\omega}{\pi} \right)^3 (A^2) \frac{t^3}{3} \Big|_0^{\pi/2\omega} \right]^{1/2} = \frac{A}{\sqrt{3}}$$

$$= 0.5774A$$

The concept of RMS voltage is introduced by some systems used to measure ac voltage. Also, the work done by a current or voltage is equal to the product $I^2R = E^2/R$. The RMS voltage (or current) represents the effective work available from a steady wave form.

The electrostatic movement has already been mentioned. It also measures true RMS voltage. Therefore, we have two systems that are capable of measuring true RMS voltage and are also capable of measuring dc voltage. Basic ac standard sources are available from several manufacturers. These sources may be purchased and sent to the NBS for calibration and certification. Then a working standard will be available for the calibration of laboratory voltage-measuring devices.

The calibrated oscilloscope can be used (with the internal sawtooth oscillator and triggering on the input signal) to obtain a standing wave form of the steady input signal. This plot can be used to obtain an equation of the input voltage as a function of time, $e_{1i} = f(t)$. If the wave form is recognized to be a sine wave, a square wave, or a triangular wave, the only values needed from the oscilloscope plot are the amplitude, A, and the frequency, ω. However, if the wave form is not one of these, it may be necessary to find the value of voltage at a number of different times and fit a polynomial to the set of ordered pairs. The oscilloscope can be used to obtain the true RMS voltage in this way.

Figure 6.43 shows the essential elements of a thermal converter. The current flowing through resistor R_2 heats the element by an amount I^2R_2. The thermocouple senses the change in temperature, and the output of a thermocouple in an evacuated tube is proportional to the I^2R_2 term. Knowing R_1 and R_2, the voltage squared can be found in terms of the thermocouple proportionality constant. Therefore, the thermal converter can be used to measure true RMS. It is a sensitive system and can measure very-high-frequency signals. High-frequency signals are difficult to obtain both with the dynamometer and the electrostatic movement. However, it is

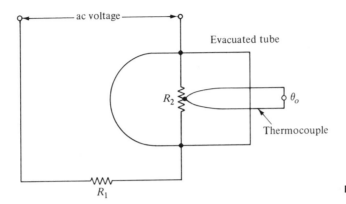

Figure 6.43 Thermal Converter

possible to burn out the thermal converter without knowing it. Care must be taken in its use.

Other types of ac voltage measurement can be made using some type of rectifier circuit to change the ac signal to a dc signal. Then the D'Arsonval-meter movement can be used to indicate the output. One very important fact must be realized when using this type of measurement system: rectifiers provide a dc signal that is the average value of the impressed voltage. For a sine wave the average value is 0.636 times the amplitude ($0.636A$). Most meters employ a constant amplification on their output scale to provide the RMS voltage. This constant is based on the sine wave (i.e., the indicated output is $0.707/636 = 1.11$ times the average value of 0.707 times the amplitude). If the input signal to these meters is a sine wave, the output is true RMS; but for any other wave form, the output is just 1.11 times the average impressed voltage. The rectifying circuits may be simply a circuit of dc storage capacitors, in which case the output dc voltage may be the amplitude only. (This is one of the least desirable types of rectifiers, since only the peak voltage is sensed. The output scale for this type of device would employ a 0.707 constant.) When diodes or electron tubes are used for half-wave or full-wave rectification, the dc voltage represents the average impressed voltage. For signal inputs with frequencies less than 10 Hz, these instruments give an output which oscillates.

True RMS meters are also available with electronic analog circuits which actually square the signal, integrate it, take the mean value, and then take the square root of this value. These systems are generally reliable for simple wave forms, but calibration is relatively meaningless for very complex wave forms.

The accuracy of voltage-measuring devices is expressed in per cent of full-scale reading. Highest-accuracy systems maintained by the NBS are used to calibrated secondary standards. Laboratory standards with the highest accuracy have errors of ± 0.01 per cent. Potentiometers for use in the laboratory may have accuracies in the range of 0.01–0.05 per cent. Secondary standards for experimental work range from 0.25 to 0.5 per cent. To understand the meaning of these accuracy figures, suppose that two systems are available. One has 100 V full scale with an accuracy of 0.25 per cent, and the second has 30 V full scale with an accuracy of 0.5 per cent. Suppose the input signal is 20 V. The 100 V meter would measure 20 ± 0.25 V, while the 30-V meter would measure 20 ± 0.15 V. Therefore, the 30-V meter would give the more accurate reading.

6.8
Resistance-
Measuring
Systems

The resistance of a mercury column 100 cm long and 1 mm in diameter maintained at 0°C was one of the first standards for resistance. This siemens unit is approximately 6 per cent smaller than the ohm. Now the possibility of defining the ohm in terms of a capacitor exists, where the only limitation on accuracy is the permittivity of space and the speed of light measurement. The NBS maintains ten 1-Ω resistors under fixed conditions. Comparison of these resistances with the defined absolute ohm indicates that the mean value of this group of resistors differs from the absolute ohm by less than one part in 10^6 (reference 6). The NBS provides calibration for secondary-standard resistors with accuracy limits. Each laboratory can maintain a set of secondary-standard resistors.

All resistors do not obey Ohm's law ($R = E/I$) when current flows through them. However, we will be concerned here with only resistors which do obey Ohm's law. Precision resistors (± 0.01 per cent accuracy or less) require protection against the environment and against oxidation. Resistors with accuracies ± 0.1 per cent do not require this protection. Resistance does change with temperature, strain, humidity, and property change (oxidation, annealing).

One basic measurement system for resistance is to measure the current flowing through the resistor and the voltage drop across it. The simple current meter of Figure 6.35 could be used to measure resistance if a known fixed voltage source were provided to be applied across the resistor. This is the method of operation of many ohmmeters. Calibration at zero resistance and at infinite resistance is easily accomplished using these meters. The accuracy depends on the accuracy of the known voltage, the current meter, and the resistance of the connecting circuitry. If a voltage-balancing potentiometer or vacuum-tube voltmeter is available, it is possible to obtain the value of an unknown resistor with an accuracy of 0.1 per cent using one standard resistor. Here the voltage source is connected to the known resistor and unknown resistor in series. The ratio of the voltage drop across each resistor is equal to the ratio of the resistors:

$$\frac{V_1}{V_2} = \frac{IR_1}{IR_2} = \frac{R_1}{R_2}$$

The only other basic method for measuring resistance that will be discussed here is the bridge method. Several different bridge methods will be covered to allow some insight into the basic problems involved in resistance measurement. The resistance bridges to be discussed here are the Wheatstone bridge, the Mueller bridge, and the Kelvin bridge (the last two are modifications of the basic Wheatstone bridge).

The basic Wheatstone bridge has already been introduced and discussed for both null and deflection operation. The only additional comments required here are the range and accuracy of operation. For resistances of 1 Ω and above, the Wheatstone bridge can measure one unknown resistance with an accuracy of 0.01 per cent. For resistances lower than this, the mechanical clamping contact resistance can be of the order of 0.001 Ω, and the bridge must be modified to eliminate this effect. The Wheatstone bridge may have two fixed legs of the bridge in the form of a ratio only. If

$$\frac{R_x}{R_1} = \frac{R_2}{R_3}$$

the ratio of R^2/R^3 may be selected constant for some bridges. Then R_1 is adjusted until null condition exists and R_x is determined.

The Mueller bridge is used for the accurate measurement of resistors less than 110 Ω. The ratio resistors (R_2 and R_3 of the above equation) are connected by a slide-wire resistor for fine balancing of the ratio. The unknown resistor is of the four-lead type (two leads are soldered to the initial resistor, one at each side of the resistance. Figure 6.44 shows the bridge. When R_v is equal to R_x, any bridge unbalance is a result of difference in resistor leads L_1 and L_4. Changing the positions of L_1, L_4, and the galvanometer to lead L_3 can detect this difference. The average value of R_x in these two bridge connections is the value R_x would have if both lead resistors were equal. This particular bridge is used very effectively with platinum-resistance thermometers. The bridge effects the change in connections by internal switching, so that change in contact resistance is not possible.

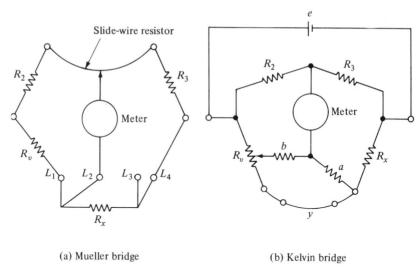

(a) Mueller bridge   (b) Kelvin bridge

Figure 6.44 Mueller and Kelvin Bridges

The Kelvin bridge of Figure 6.44 is used for measuring resistances less than 1 Ω. The slide-wire variable resistor, R_v, and the unknown resistor, R_x, both have four lead connectors. The value of resistance is

$$R_x = R_v \frac{R_3}{R_2} + \frac{b(R_3/R_2) - a/b}{a + b + y}(y)$$

The slider, R_v, is usually a manganin bar with sliding contact. The effect of sliding contact and contact resistance is essentially eliminated. If $ab = R_3/R_2$, the output is the same as that of the ordinary Wheatstone bridge circuit. Well-designed Kelvin bridges (double bridges) can measure resistance from 1 Ω to 10^{-8} Ω and perhaps lower.

6.9
Conclusions

The standards for the measurement of length, time, and mass provide accuracies in the range of one part in 10^8 or 10^9 of the standard unit of measure. Many of the extraneous inputs that influence the measurement of these dimensions have been explicitly pointed out. Since all other dimensions do depend on these three defined units, the accuracies of all other dimensions are limited by the accuracy with which these three dimensions can be determined. This chapter has emphasized the limitations of measurement of primary dimensions so that an understanding of the whole field of measurements will be clearly established. The logical development of measurement systems is based on the principles set forth here.

Once these three standards were well defined and understood, it was possible to introduce the concepts of standards traceable to these standards by physical laws. It was also possible to explain the difficulty involved in the definition of new measurement standards in order that a consistent set of measurement standards exist. The problem of defining temperature, voltage, and resistance standards that would be consistent with both the previously defined standards (time, length, and mass) and with the physical laws was discussed. The logic involved in these methods of establishing consistency is the same as that required for any new dimensions that will be defined in later chapters. It is important, therefore, to approach all new problems with the same logic.

The length of space required to represent properly the conditions imposed on derived standards is indicative of the complexity of the problem. In later chapters the details involved in this comparative process will be somewhat shortened, but it must be realized that all discussions in this chapter are also applicable to any measurement system that might be considered. All future measurement systems will be influenced by the same extraneous inputs as those discussed here. Therefore, it is not necessary to repeat the precautions that have already been introduced. In the presentation of different types of measurement systems in Chapters 7–9, and liberal use of the material in this chapter will be made.

The basic concepts of measurement have been presented in this chapter. The first object in the design of a system to measure any dimension is to obtain a sensor capable of detecting changes in the dimension to be measured. Then, an understanding of the sensor response to all types of extraneous inputs must be obtained. Finally, the system must be calibrated by using a standard known input over the range of values of the dimension to be measured. After calibration, the measurement system is ready for use in determining the value of the unknown signal. However, it is still important to evaluate any sources of extraneous input that are present in the measurement process. It is possible to calibrate and evaluate extraneous inputs by considering the total measurement system (sensor, intermediate modifying device, and read-out device). For this method, the output of the total system is compared to the standard input signal, and the effect of extraneous inputs on the output signal is considered in one operation. If this technique is employed, it is possible to eliminate most of the sources of errors in measurement. The absolute elimination of errors is not possible, since the very act of measuring changes the system being measured and complete evaluation of extraneous inputs requires a knowledge of the position and velocity vector of all masses in the universe. Fortunately, the accuracy capabilities of a measurement system are seldom limited by the lack of an absolute elimination of errors. Therefore, an analysis of the change in the system being measured and the major extraneous inputs can be completed to the point where the errors not evaluated do not constitute a significant part of the measurement error. The measurement system can then function properly to obtain values of the dimension being measured.

References

1. *National Bureau of Standards Technical News Bulletin*, vol. 52, no. 1, U.S. Department of Commerce, Washington, D.C., Jan. 1968.

2. SWINDELLS, J. F. (ed.), *Precision Measurement and Calibration—Temperature*, NBS Special Publication 300, vol. 2, U.S. Department of Commerce, Washington, D.C., 1968.

3. BECKWITH, T.G., and BUCK, N.L., *Mechanical Measurements*, 2nd ed., Addison-Wesley Publishing Co., Reading, Mass., 1969.

4. HAMMOND, D.L., and BENJAMISON, A., "Linear Quartz Thermometer," *Instruments and Control Systems*, Oct. 1965, pp. 115–119.

5. *Handbook of Electrical Measurements*, "Instruments and Control Systems," Chilton Co., Rimbach Publication Division, Pittsburgh, Pa., 1966.

6. *Precision Measurement and Calibration, Electricity—Low Frequency*, NBS Special Publication 300, vol. 3, U.S. Department of Commerce, Washington, D.C., 1968.

7. *Precision Measurement and Calibration, Electricity and Electronics*, NBS Handbook 77, vol. 1, U.S. Department of Commerce, Washington, D.C., 1961.

PROBLEMS

6-1. The accuracy with which a standard can be defined limits the accuracy with which a dimension can be measured. What is the maximum accuracy available for measuring length, time, and mass?

6-2. Why can a drafting scale be a standard? To what extent is this standard valid?

6-3. Select any ruler and determine the maximum accuracy obtainable with this device. Do the same for a micrometer. What physical property limits the accuracies of these devices?

6-4. If length, mass, time, and force were all given primary standards for measurements, what constraints would have to be satisfied?

6-5. How would volume be determined from the standards given in this chapter? Give one physical law that could be used to provide a sensor for measuring volume.

6-6. A surveyor's transit is to be used to measure the distance between the transit and the rod. When the distance is 100 ft, the operator reads the angle between the horizontal and 1 ft on the rod below the horizontal. Determine the sensitivity (the change in the angle per change in distance between transit and rod, $d\theta/dL$) of this measurement system.

6-7. Gage blocks with an accuracy of $\pm 2 \times 10^{-5}$ in. are used to measure an unknown length (see Figure 6.1). Four gage blocks are required to obtain the correct dimension. What is the accuracy of this measurement?

6-8. A micrometer is calibrated and accurate at 70°F. The distance between the moving shaft and the fixed head (Figure 6.2) is 2 in. when the shaft is fully open. When the temperature is −330°F, what thermal error is possible in the measurement of a 1-in. dimension? (Assume that the linear coefficient of the micrometer material is 2×10^{-5} in./in.°F).

6-9. The ballast circuit of Figure 6.13 has a current-measuring meter of zero resistance inserted in the circuit. Find the sensitivity of this current-sensitive device (di/dR_s).

6-10. The Wheatstone bridge circuit of Figure 6.14 uses a voltage meter to measure e_o such that no current flows through the meter. Initially, all of the resistors have the same resistance, R_0. If only one resistance changes, what is the sensitivity of this measurement system (de_o/dR)?

6-11. How does the lead-wire resistance between the voltage source, e_i, and the points A and C (Figure 6.14) affect the accuracy of the bridge circuit?

6-12. If a differential balance is used for initial balance with a bridge having four resistances changing in the basic Wheatstone bridge circuit, use Equation 6.28 to show the effect of these parallel resistors in the bridge circuit.

6-13. A circular cross-section wire is used for a resistor. The wire obeys Equation 6.30. Find the per cent change in resistance per change in temperature $(100/R)(dR/dT)$ if the linear coefficient of expansion is 2×10^{-5} in./in.°F and the resistivity does not change with temperature.

6-14. A mass balance can measure with an accuracy of ± 0.001 g. What is the per cent accuracy in measuring 1 kg?

6-15. A counter with five digits is calibrated using station WWV. What is the maximum accuracy with which this device can measure?

6-16. A counter with eight digits is calibrated using a very stable oscillator, an oscilloscope, and station WWV. If the system can be used to calibrate as accurately as the broadcast signal can be received, what is the maximum accuracy of the calibrated counter?

6-17. The oxygen point, the triple point, the steam point, the zinc point, and the sulfur point are used to obtain the constants, R_0, A, and B of Equation 6.36. The corresponding resistances are 100 Ω, 118 Ω, 130 Ω, 170 Ω, and 200 Ω. Find the best values of the constants by a least-squares technique.

6-18. The reference temperature for a copper–constantan thermocouple circuit (where the iron of Figure 6.30, A, is replaced by constantan) is at 20°F. If the measuring junction is at 200°F, what is the output of a standard thermocouple?

6-19. For the copper–constantan thermocouples of Figure 6.29(a) (constantan for iron) what is e_o if $t_1 = 20°F$, $t_2 = 40°F$, $t_3 = 60°F$, and $t_4 = 80°F$?

6-20. If constantan replaces iron in Figure 6.29(b) and each junction at t_2 is at 200°F while T_1 is at 100°F, what is the output voltage? If the T_2 junctions have values of 200°F, 160°F, and 140°F, what is the output voltage?

6-21. Fit the least-squares straight line of emf $= A + BT$ through the points of Table 6-2 for copper–constantan thermocouples for temperatures 20, 40, 60, 80, and 100°C. Find the standard deviation of the points from this curve.

6-22. Repeat Problem 6-21 for a parabolic, emf $= A + BT + CT^2$.

6-23. When a voltage-balancing potentiometer is used to measure voltage, null balance occurs when zero current flows through the meter. A potentiometer with accuracy of ± 0.01 mV is used in a circuit. If the circuit resistance in the input voltage–galvanometer–slide-wire circuit is 2000 Ω, what is the maximum possible current that could flow through this circuit under null conditions determined by the galvanometer?

6-24. If the oscilloscope is used in x–y plotter operation, $x = A \cos 20\pi t$, and $y = B \cos 30\pi t$, what will be the scope output?

6-25. What is the RMS voltage of the periodic wave

$$V = At \qquad 0 \leq t < \frac{\pi}{4}$$

$$V = A \qquad \frac{\pi}{4} \leq t < \frac{3\pi}{4}$$

$$V = 4A\left(1 - \frac{t}{\pi}\right) \qquad \frac{3\pi}{4} \leq t < \pi$$

7
DERIVED DIMENSIONAL
QUANTITIES

7.1
Introduction   With the establishment of standards for time, length, mass, temperature, resistance, and voltage, it is now possible to use the physical laws and definitions to derive standards for the measurement of any other dimensional quantity. The present chapter is concerned with dimensional quantities which are simply related to the basic dimensions. The dimensions concerned with energy and mass transport are considered in Chapter 8, since they involve more complex relationships between the basic defined standards.

The simple derived dimensional quantities are considered in the order length-type dimensions, length- and mass-type dimensions, length- and force-type dimensions, length- and time-type dimensions, mass- and time-type dimensions, time- and temperature-type dimensions, and electrical-type dimensions. This order was selected to present the progressive development of a dimensional system. It is one goal of this chapter to logically outline the techniques of developing any dimension from the basic defined standards. The reader should keep this goal in mind when reading the chapter so that the application of these methods to any dimensional quantity can be accomplished.

Certainly, a good physical understanding of any dimension is a prerequisite to the proper application of any basic set of standards in the development of a derived standard. It is always more fundamental to base all the aspects of the derived standard on the six standards defined in Chapter 6. Since all of these standards are well defined and offer the greatest accuracy in measurement, any dimension based on these standards will have the maximum possible accuracy when defined by these basic standards. However, it is almost always possible in practice to use derived standards to define new standards with the same accuracy as that obtained by definition from the basic standards. In the early sections of this chapter, the maximum accuracy of the derived standards will be discussed to acquaint the reader with the limitations that are present in any defined dimension. Later sections will dispense with the formality of examining these accuracy considerations, with the understanding that this type of analysis must be performed to understand completely the limits on the derived dimension.

The standards derived in this chapter will be building blocks for the development of standards for any measurement of any dimensional quantity. They should be used with more caution for one major reason: derived standards must be less accurate than the basic dimensions defined in Chapter 6. It would not be good engineering practice to develop standards for the measurement of volume and area, and then to use these defined standards to obtain a method for measuring length (dividing volume by area). One should always base measurements on the best standard available. In any case, a simple consideration of the maximum possible accuracy obtainable should be made. It is possible that in some laboratory which has only limited equipment the measurement of a subtended angle may be the most accurate method available for measuring length. One should not hesitate to use the most accurate measurement system available just because its defining standard was derived from more basic standards. However, when maximum accuracy is the dominant requirement of a measurement system, the use of basic standards should generally be incorporated in the development.

**7.2
Geometric
Length
Considerations**

The development of the measurement of angles, area, and volume is based on geometric relationships. Angle blocks are available certified by the NBS in the form of steel wedges with a fixed angle between the surfaces of the wedge. These blocks are used in the same manner as gage blocks. These are the practical working standards for angular measurement. Their accuracy is $\pm 0.01''$ of arc certified by the NBS. However, angles are fundamentally defined by the length standard and geometry. The geometric construction of an equilateral triangle using standard lengths can be used to generate a 60° angle. This can be divided by bisecting the angle as many times as necessary. The use of a laser beam can provide extremely precise determination of an angular displacement for static calibration. Figure

7.1 gives one possible arrangement for this calibration. The measurement of x and y, and the geometric relationship between radius and the cord of an arc, can be used to determine the angle. The highly collimated laser beam will provide two index marks for the determination of y. To measure the angle requires two length measurements, and the error is twice that for one measured length.

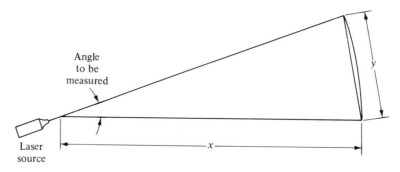

Figure 7.1 Laser Calibration of Angles

Geometry and the laser beam can also be used in the arrangement of Figure 7.2 to determine the angular orientation of a specularly reflecting mirror. Here the optical square is determined by first obtaining the mirror position (or laser position) where the reflected ray returns directly to the source. Then the angular displacement from this orientation can be measured on the scale in the same manner as in the previous description. Here the length along the circular scale and the radius of the circle can be used to find the angle: scale length $= R\theta$, $d\theta = dL/R - (L/R^2)\,dR$.

A third method for measuring angular position is the use of the circularly wound resistance with slider of Figure 7.3. Here the variation of resistance with

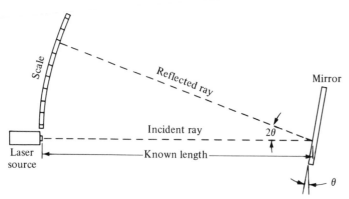

Figure 7.2 Laser on a Spectral Mirror

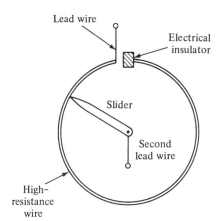

Figure 7.3 Resistance Type Angular Sensor

angular position results from the expression for electrical-resistance variation with length, $R = \rho L/A$, where ρ and A are constant. The resistance varies linearly with length along the wire and is directly related to the angular position of the slider, $dL = A/\rho \, dR$. The error in measuring the length and radius gives the error in angle. All systems may be calibrated using angle blocks or by using a standard length-measuring system and geometry. The technique of unwinding a string from a circular spool of measured diameter is shown in Figure 7.4. This transforms an angular rotation into translational motion. The errors are the same as before.

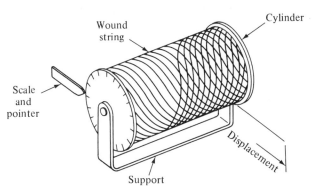

Figure 7.4 Angular Displacement Calibration

AREA

The measurement of area in a plane is also possible using the standard length and geometry. Area is defined for a perfect square to be the product of the length of two of its equal sides. A rectangle can always be divided into groups of squares. Geometry and calculus can be used to determine any real area if a sufficient number of length measurements are made. Many measurement systems based on geometry are available for the measurement of area. Some are for special shapes, while others may be applied to any closed area. The polar planimeter is perhaps the most commonly used area-measuring system capable of measuring any closed planar area.

The essential elements and dimensions of a polar planimeter are shown in Figure 7.5. The knife-edge wheel rotates only when movement results

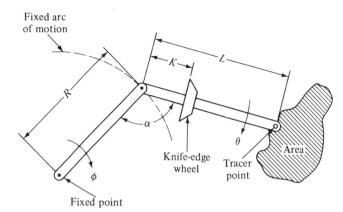

Figure 7.5 Elements of a Polar Planimeter

in an angular displacement of θ. From geometry it can be shown that the total area (neglecting higher-order differentials) swept out by both arms as seen by the wheel of the planimeter is composed of three parts:

$$\text{area by arm } R = \tfrac{1}{2} R^2 \int d\phi$$

$$\text{area by arm } L = \text{area of rotation} + \text{area of translation}$$

$$\text{area of rotation} = \tfrac{1}{2} L^2 \int d\theta + \tfrac{1}{2} L^2 \int d\phi - L \int K \, d\phi$$

$$\text{area of translation} = LR \int \sin \alpha \, d\phi$$

Total differential area is:

$$dA = \tfrac{1}{2} R^2 \, d\phi + \tfrac{1}{2} L^2 (d\theta + d\phi) - LK \, d\phi + LR \sin \alpha \, d\phi$$

The distance traveled by the wheel in this process is given by

$$dD = K \, d\theta + R \sin \alpha \, d\phi$$

If the tracing point moves around the area back to the original starting point, $\int d\theta = 0$; and when the fixed point is outside the area being measured, $\int d\phi = 0$. Then

$$A = \tfrac{1}{2}L^2 \int d\theta + (\tfrac{1}{2}L^2 + \tfrac{1}{2}R^2 - LK) \int d\phi + RL \int \sin \alpha \, d\phi$$

$$= RL \int \sin \alpha \, d\phi$$

and

$$D = K \int d\theta + R \int \sin \alpha \, d\phi = R \int \sin \alpha \, d\phi$$

Therefore, $A = LD$. The distance traveled by the circumference of the wheel can be measured by an angular scale and rotational counter for a fixed diameter. The lengths of the arms are fixed and known, so the area is determined. The error in measuring A is $dA = L \, dD + D \, dL$, which is twice the error in measuring length.

When the fixed point is located inside the area being measured, $\int d\phi = 2\pi$; and the area is given by

$$A = LD + (R^2 + L^2 - 2KL)(\pi) \tag{7.1}$$

Calibration is easily accomplished by constructing a perfect square with measured sides or a perfect circle of measured radius. In either case geometry and a standard length dimension provide the standard for area calibration. To avoid backlash (or lost motion) in the wheel, the tracer point should be moved along the area to be traced and stopped at the starting point to record the scale reading. Then the trace can continue until the starting point is reached again.

The second term on the right side of Equation 7.1 represents the area of the zero circle of the planimeter. This circle is the circle that would produce no net rotation of the wheel if the fixed point were placed inside it. This term can be found experimentally by constructing two circles with areas different from the zero circle and measuring the areas with the planimeter. The radius of the zero, R, circle can be found with the fixed point inside the areas by using Equation 7.1 where the area is known and the reading LD is measured. Then

$$A_1 = LD_1 + \pi R^2$$
$$A_2 = LD_2 + \pi R^2$$

and

$$2R^2 = A_1 + A_2 - L(D_1 + D_2)$$

The use of two or more circles is recommended to average the measurement errors.

SURFACE AREA

When the area to be measured does not lie in a plane, the curvature of the surface and the corresponding three-dimensional nature of the problem cause serious measurement difficulties. It is possible to mold a thin plastic film over the surface, trim it to the exact dimensions, and cut it in enough places that it can be placed in a plane. Then the planimeter can be used to determine the surface area. A second possibility is to find the equation of the surface to be measured in the form $z = f(x, y)$. Then the evaluation of a double integral of the area projected into the x–y plane could be completed. If the function cannot be found analytically, a numerical solution obtained by finite difference methods is another possibility. The errors involved in this method can be quite large even when the values of z are measured at selveral hundred x–y points. The problems of finding the perpendicular and measuring enough points must be solved with small errors. The method of using a thin plastic sheet molded on the surface certainly deserves consideration in this problem.

SURFACE ROUGHNESS

The measurement of surface roughness is a special case of the above measurement problem. In this situation the surface being considered is generally flat, and the irregularities in the surface may be of the size of 1 μin. However, many modern problems remain unsolved because it has been impossible to describe adequately the condition of the surface. (Nucleation sites in boiling, electrical contact, and radiation properties of surfaces are a few of the areas needing more adequate surface descriptions.) The present standards for comparing surfaces incorporate several flat plates polished with various grit-size powders. The surface of a polished standard is still irregular, but the size of the irregularities is somewhat controlled. A typical surface might look like that shown in Figure 7.6. The distance between the peaks and pits is frequently called the peak-to-peak displacement. The surface line can be defined in many ways. One way is

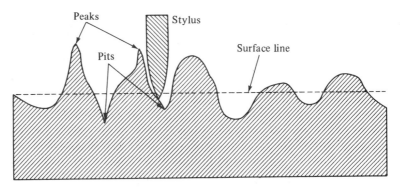

Figure 7.6 Magnified View of a Typical Surface

to require that the area of the material above the line be equal to the void area below the line. The distance from the peaks to the pits may be only 1 μin. for polished surfaces, with a similar dimension between peaks. This measurement problem has not been adequately solved. Two frequently used methods of measuring surface conditions will be presented and several possible measurement systems will be suggested.

A stylus similar to the one shown in Figure 7.6 is used to detect the vertical displacement as it moves across the surface. This stylus is exactly the same as a phonograph needle, but the output of the stylus may be sensed by several types of intermediate sensor systems. The stylus has three major problems. First, from the figure it can be seen that the size of the stylus is too large to reach the bottom of the pit where it is located. When the stylus is moved across the surface, it destroys the peaks by bending and wearing the peaks. Finally, the stylus only measures one line of the surface with line width equal to the stylus diameter. Decreasing the stylus diameter allows it to penetrate farther into the pit, but it also does more damage to the surface since the contact force is concentrated at a smaller point. The common solution to the problem of sampling only a line of the surface is to assume that the surface is uniform in all directions. None of these techniques is totally satisfactory for an adequate surface description. Most of the measurement problems that have not been solved today involve the interface between two different materials. An understanding of the physics of the problem awaits the development of a measurement system capable of describing the surface adequately.

The output of the stylus may be sensed by a piezoelectric crystal, a variable inductor, or some other variable-electrical-property sensor. The electrical output is usually amplified, and the RMS value of the surface roughness is the final output. Calibration is accomplished using prepared polished plates. The most common form of the output is in microinches RMS. This type of system is probably the most commonly used surface-roughness measurement system. It is certainly more descriptive than examining the surface using the finger, but it is equally certain that it will never completely describe the surface. In some cases the same instrument in used to obtain the RMS value of the slope of the surface; sometimes better correlation of a physical phenomenon results using this information. The errors involved in this measurement have not been evaluated.

A second method used in the measurement of surface is the focusing microscope. With this system it is possible to focus the microscope first on the peak and then on the pit. The difference between these vertical positions is determined from the change in focal length. Three major problems are also associated with this measurement system. The area sampled is much smaller than that sampled by the stylus system, the operator time and the precision obtainable are unsatisfactory, and the maximum pit depth is limited by the amount of light and the incident angle that can be used on the sample area. When the magnification is increased, the light required also increases. Visual inspection is time consuming, and nothing is learned about the massive amount of area not viewed. Still, for fairly uniform surfaces the measurement has its uses.

Several other methods are possible in the measurement of surface roughness. Four methods will be mentioned at this point. The use of a laser source for making holographic pictures of a surface offers one possible solution. The holographic picture is a three-dimensional picture of the surface and requires visual interpretation of the output. However, the representation can be calibrated by introducing a known dimensional axis at the surface. A second method is to immerse an electrically heated plate in pure water. If the air is essentially removed from surface cavities by prolonged boiling, the initiation of nucleate boiling sites should be a function of the temperature of the plate and the size of the cavity on which vapor bubbles appear. Again, visual observation is required and the cavity size is only a relative measure unless cavities of known dimension can be made in the surface. Another comparative method is to test a plate in an electrolytic fluid adjacent to a fixed plate. The electrical capacitance depends on the true surface area of the plate. It might be possible to vapor plate a smooth glass surface to be used as the standard surface for comparison. The capacitance would increase with surface roughness, since the rougher surface would have a larger surface area. A final possibility to be discussed here is the use of the wave nature of light to describe a surface. When the wavelength of light is larger than the surface cavities, the surface should reflect light specularly (angle of incidence equals angle of reflection). Using a collimated monochromatic light source, the wavelength can be decreased until photosensors indicate that the light is no longer reflected specularly. This occurs when the wavelength is small enough to enter the cavity, and with multiple reflections, the light can leave the surface at any angle.

VOLUME

Volume can also be defined using geometry and a standard length dimension. The volume of a perfect cube is equal to the length of a side cubed ($dV/V = 3\,dx/x$). The volume inside and outside of a container is defined by this relationship. Also, a perfect sphere could be used with the same result. The volume inside or outside a complex geometry can be measured by the methods to be introduced in the following section. The ideal gas relationship can be used to measure the volume of a container in the following manner. The volume to be determined can be evacuated to the point where essentially no mass is left. A second container of known volume (one perfect cube) with measured temperature and pressure can be connected to it. Then the valve between these chambers can be opened, and the final temperature and pressure can be measured. The final volume is determined from

$$\frac{P_1 V_1}{T_1} = \frac{P_2(V_u + V_1)}{T_2} \tag{7.2}$$

where V_u is the unknown volume and is the only quantity not known in the above expression. This would require that the system have zero heat transfer across the walls. The use of an ideal gas introduces some error into the system; but for moderate temperatures and pressures, air can be used with little error. The errors in measuring temperature and pressure would probably dominate the measurement accuracy. The temperature error has already been discussed in Chapter 6. Pressure errors will be considered later.

7.3
Mass and Length Combinations

The measurement of density (or specific volume) involves the use of two basic standards in establishing a measurement standard. The most straightforward method of determining the density of a liquid is to balance a container of known volume (say 1 ft³), fill the container with the liquid, and balance the mass of the liquid added to obtain the density (mass divided

by 1 ft³). A variety of other methods is given in reference 1. Since density
and volume vary with temperature, this measurement should be made for
a range of temperatures and the corrections should be added. If the buoyancy of the fluid in air is significant, this correction should also be applied.
Thermodynamic tables for the properties of several liquids have been made
based on measurements of this type. The steam tables give the density
of water with an accuracy of approximately one part in 6000. If this accuracy is acceptable, water can be used to measure the volume of an irregularly
shaped container. In this mode of operation, the density is known and the
mass is measured. The volume is the mass divided by the density. Temperature must also be measured.

Once the density of a liquid is known as a function of temperature,
the liquid can be used to place etched marks on a glass buret. Very accurate
volumes can be obtained by building burets with the internal geometry of
Figure 7.7. The very thin tubes on each end of the bulb allow for sensitive

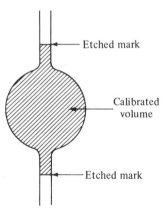

Etched mark

Calibrated
volume

Etched mark

Figure 7.7 Schematic of a Buret

control of the volume contained. The density of a liquid is a very appropriate laboratory working standard. Impurities in the fluid are extraneous
inputs into the system. Liquid densities can be measured accurately with
the secondary standard volume and the mass balance. The basic error in
density is given by $d\rho = dm/V - m\,dV/V^2$ or $d\rho/\rho = dm/m - dV/V$.
The error in measuring volume can then be used to obtain the error in
measuring density $(d\rho/\rho = dm/m - 3\,dx/x)$ in terms of basic dimensions.
The meniscus formed between air and the fluid being investigated must be
evaluated for precision calibration.

The measurement of gas density can be accomplished in a similar manner. A chamber of known volume is evacuated and the mass determined.

The gas of interest is introduced into the volume and the mass determined again. It is particularly important to note that the buoyancy of air need not be considered if the volume was closed and evacuated in the initial mass balance and only the gas being considered is allowed to enter the volume.

The density of a solid is obtained by measuring the volume of the solid and then performing a mass balance. If the volume configuration is too complex for simple geometry, it can be determined by immersing the solid in pure water at a known temperature. A measurement of the solid mass submersed in the water is less than the mass measured in air by the buoyancy of the volume of water displaced. (Neglecting the density of air compared to that of water.)

$$m_{air} = m_{water} + V_{solid} \rho_{water} \tag{7.3}$$

or

$$V_{solid} = \frac{m_{air} - m_{water}}{\rho_{water}} \tag{7.4}$$

If the solid has very small cavities in it, the water may not completely fill these cavities and an error results. If the errors in measuring all quantities (m_{air}, m_{water}, and ρ_{water}) can be estimated, the overall error in measuring volume can be approximated by

$$\Delta V_{solid} = \frac{\Delta m_{air} - \Delta m_{water}}{\rho_{water}} - \frac{(m_{air} - m_{water})}{(\rho_{water})^2} (\Delta \rho_{water})$$

The density of the solid can now be determined to be

$$\rho = \frac{m_{air}}{V_{solid}}$$

Another form of presenting density data is to express the ratio of the density of any substance to the density of water at a given temperature. This ratio is called the specific gravity, S:

$$S = \frac{\rho}{\rho_{water}}$$

The density of many solids, liquids, and gases has been determined and is available in tabular form.

If a solid is soluble in water, some other fluid with known density must be used. The section on mass measurement in air should be referred to in any mass measurement. When the density of the mass being measured is approximately equal to the density of the standard masses, the effect of air density is negligible.

7.4
Force- and Length-Type Systems

The dimension, force, is determined from the standards for time, length, and mass and from Newton's physical law of acceleration:

$$F \propto \frac{d(mv)}{dt} = kma$$

where F = force, m = mass, v = velocity, t = time, a = acceleration, and k = the constant of proportionality. The vector direction can be determined from geometry. A great deal of misunderstanding surrounds the dimensions for force, mass, and this constant. One reason for this misunderstanding is the fact that two methods are available for defining force. If force is defined in terms of previous standards to be 1 lb_m-ft/s², the constant of proportionality is dimensionally and numerically equal to 1. The dimensional units of force would then be lb_m-ft/s². Then the local accleration of gravity at sea level would attract 1 lb_m with a force equal to 32.174 lb_m-ft/s² (980.665 dyne). However, the pound-mass system of measure defines 1 lb_f to be the force that would accelerate 1 lb_m at the rate of 32.174 ft/s². The constant of proportionality can be found to be

$$K = \frac{F}{ma} = \frac{1\ lb_f}{32.174\ lb_m\text{-ft/s}^2} = \frac{lb_f\text{-s}^2}{32.174\ lb_m\text{-ft}}$$

For the dimensional system in which force is defined independently of the basic dimensions, the constant K is a constraint on the dimensional system. This constant is dimensionally and numerically equal to 1 (i.e., $1 = lb_f\text{-s}^2/32.174\ lb_m\text{-ft}$). This constant may be multiplied by any dimension in this dimensional system without changing the value of the quantity. Since there are four commonly used definitions of dimensional systems, it is not surprising that the complete understanding of dimensions requires some experience.

The conflicts that arise from the arbitrary definition of a dimension that has already been defined by a set of standards and physical laws is apparent from the previous paragraph. Fortunately, this practice has been limited. From a measurements point of view, it would be more logical to have the units of force be lb_m-ft/s². However, the proportionality constant has already been evaluated, and the standard for force can be derived directly from the basic standards. By measuring the local acceleration of gravity (to be discussed in the next section), the simple mass balance is transformed into a force balance; and one force-measuring system is available for direct use. Any physical law relating force to other dimensions can be used to develop a force-measuring system. The systems of Examples 3-1–3-5 have already been discussed in detail. They should be reviewed with respect to the standards for length, mass, and time. Any of the physical laws of Example 3-1 can be used for measuring force. Two types of force-measuring systems will be discussed here: the lever type and the spring type.

All lever-type measuring systems have two characteristics in common. They actually are mass-comparing systems requiring the local acceleration of gravity to provide a value for force. They also can have a slight shift in the center of mass when operating in a displaced position (deflection operation). The accurate evaluation of the local acceleration of gravity has already been mentioned as a prerequisite for force measurement. Figure 7.8

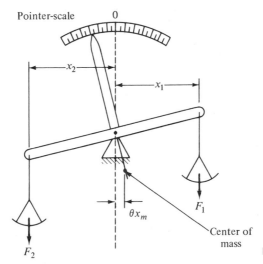

Figure 7.8 Deflection Balance

shows the simple balance in deflection operation. The center of mass shifts away from the vertical line containing the knife edge, and the total force balance is given by $F_2 x_2 = F_1 x_1 + g/g_c M_{total} x_m$. The magnitude of the total mass, M_{total}, changes with the values of the forces F_1 and F_2. One method of eliminating this input is to balance the force to be measured so that the center of mass lies in the vertical line of the knife edge. This corresponds to the null or zero pointer position. Another commonly used lever system is the platform scales of Figure 7.9. If the dimensions between the

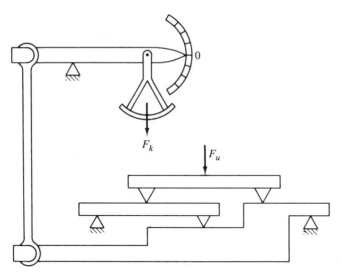

Figure 7.9 Platform Scales

knife edges are properly selected, the unknown force, F_u, can be measured at any position on the top lever with null operation. Any errors in the knife-edge locations can cause errors in the output force. Zero balance (with $F_u = 0$) can be obtained by providing a counter-balancing mass on a screw. The summation of the moments around all fixed knife edges and the summation of forces will give the dimensions required for the knife-edge positions. The complex geometry involved in deflection operation makes this lever system a null measuring system.

The spring-type force-measuring system does not depend on the local acceleration of gravity to measure force. This type of system is very broad, and a large variety of systems based on the elastic properties of materials are in common use. The common bathroom scale is one example. If the output of the platform scales of Figure 7.9 is applied to a spring-type detector, the bathroom scale results (see Figure 7.10). This type of system is of the displacement type, and the error in shifting center of mass is present. The simple spring-type system of Figure 7.11 can be used to eliminate the center-of-mass problem. In all cases the displacement of the sensor is proportional to the force applied. Reference 1 presents some of the basic considerations that must be met for force measurement. The sensors of Figure 7.11 lead directly to the concepts of stress and strain. The method of measuring lengths x and $x + \Delta x$, can be made using secondary standards for length measurement. The error in measuring force for Parts (a) and (b) of Figure 7.11 can be obtained from the relationship

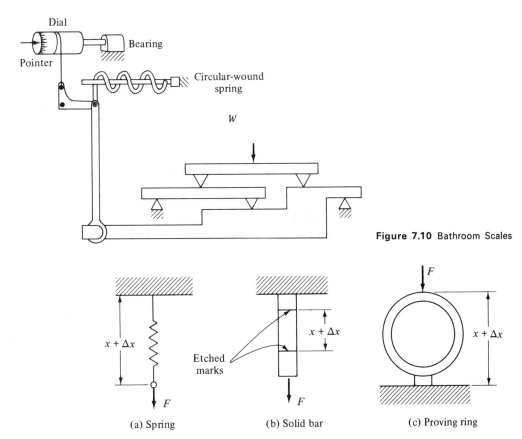

Figure 7.10 Bathroom Scales

Figure 7.11 Elastic Type Force Measurement

$$F = (\text{const})(x)$$

$$dF = (\text{const})\,dx$$

The error in force measurement with this system is also dependent on temperature and deviations from the assumed relationship. These deviations are a part of the field of stress–strain concepts.

STRESS AND STRAIN

Stress and strain are studied in the mechanics of materials. First we will consider stress–strain problems in the area of solid materials. The application to fluids will be considered in a later section, where the dimension viscosity is introduced. All solids experience a deformation of physical shape when subjected to an external force. Stress has been defined to be the ratio of the applied external force divided by the cross-sectional area perpendicular to the direction of the force. For simple geometries the measurement of area and force constitute a measurement of stress. Figure 7.11(b) is a simple geometry. However, parts (a) and (c) require additional concepts involved in the study of statics of solids. This book will discuss only simple geometric configurations, with the understanding that the methods developed can be applied to more complex systems.

Strain is defined to be a change in length divided by the original length (for Figure 7.11(b), strain $= \epsilon = \Delta x/x$). The measurement of strain is accomplished by determining two lengths. Robert Hooke first made the observation that the strain in many materials is directly proportional to the stress applied $[S = (\text{const})\epsilon]$. The constant of proportionality is called the modulus of elasticity (Young's modulus, E). Then stress, S, is given by

$$S = E\epsilon \qquad (7.5)$$

Using the simple experiment of Figure 7.11(b) Young's modulus has been determined for many materials. Three points of interest will be mentioned before the measurement of strain is developed for the determination of stress. First, the modulus of elasticity varies with temperature, material, and the type of stress applied. The first two of these parameters can be controlled to obtain a tabular listing of modulus of elasticity. The last must also consider the shear modulus of elasticity (angular strain, $\Delta\theta/\theta$, times the shear modulus of elasticity, G, equals the torque, T).

A second point of interest is that the cross-sectional dimension of a material being stressed also changes. The ratio of the strain in the transverse direction to the strain in the axial direction is called Poisson's ratio. For the member of Figure 7.11(b), the diameter perpendicular to the direction of the applied force is called D. Then the traverse strain is dD/D; and Poisson's ratio, μ, is

$$\mu = \frac{dD/D}{dx/x} \qquad (7.6)$$

The true stress must be based on the true cross-sectional area and the applied force. Frequently, data is presented for nominal stress, which is based on the unstrained diameter. This point should be clear before an analysis is started.

Finally, it should be noted that materials obey Hooke's law over only a limited range of stress application. Figure 7.12 shows the plot of stress–strain diagrams for two different types of materials. Hooke's law holds only for that portion of the curve where strain varies linearly with stress. Hooke's law holds in part (a) up to the elastic limit. When yielding occurs, a permanent elongation of the material results. Hooke's law does not exactly hold for the material of part (b), although an approximate number for Young's modulus may be listed in tables.

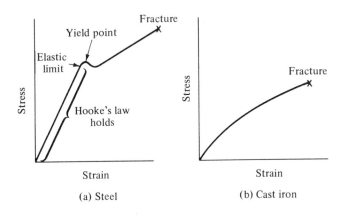

(a) Steel

(b) Cast iron

Figure 7.12 Two Typical Stress-Strain Diagrams

STRAIN-MEASURING SYSTEMS

One of the first methods for measuring strain was to place two indentations in the surface of the element being considered. Then a caliper and micrometer could be used to measure the distance between the dents. Figure 7.13 shows this method for measuring strain. Although this method has

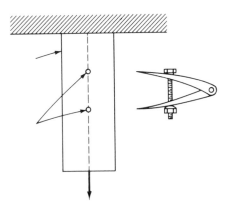

Figure 7.13 Caliper for Strain Measurement

generally been replaced with new techniques, it is still a relatively accurate method of measuring strain. Some error exists in the change in dent size with stress and in the transferring of the length from the dents to the micrometer. If a force of 100 lb is applied to a shaft with area of $\frac{1}{3}$ in.^{2}, a steel shaft with $E = 30 \times 10^6$ psi will have a strain equal to 10^{-5} in./in. or 10^{-5} cm/cm ($\epsilon = F/A/E$). The accuracy of the micrometer is generally not accurate enough to measure a 0.00001-in. change in length. This would be necessary if the initial distance between the indentations is equal to 1 in. This system is accurate enough to obtain data for a plot similar to Figure 7.12(a). Perhaps the decline of this type of strain-measurement system was caused by the need for an operator and static conditions for data collection.

A second method for measuring strain is shown in Figure 7.14. The

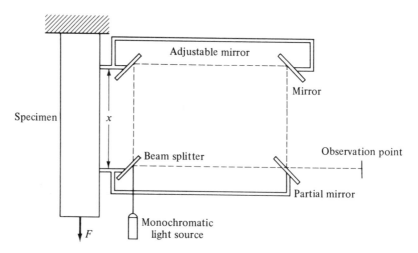

Figure 7.14 Optical Strain Measurement

mirror system and a monochromatic light source is first adjusted to a value of x for which the observer sees constructive interference (brightest spot). The length x is measured, and the force is applied. With the adjustable mirror and the beam splitter attached to the specimen, a change in x equal to $\frac{1}{4}$ wavelength will produce destructive interference (dark spot), since the distance traveled by the split beam will be $2(\Delta x)$ or $\frac{1}{2}$ wavelength. The greatest error in this system is the measurement of the initial geometry. A photomultiplier with a digital counter can be used to count the fringe patterns. The error in measuring x over some ranges is less severe than errors in measuring Δx, since

$$\frac{d\epsilon}{\epsilon} = \frac{d(\Delta x)}{\Delta x} - \frac{dx}{x}$$

The error in measuring Δx is approximately $\frac{1}{4}$ wavelength (or approximately $\pm 5 \, \mu$ in.). If the error in measuring x is ± 0.01 with $x = 1$ in., the error in measuring Δx will be the dominant error until Δx has a nominal value of $\pm 500 \, \mu$ in. For higher strains the error in measuring x dominates. The use of gage blocks should easily reduce errors in measuring x to the range of $\pm 10^{-4}$ in.

This system is cumbersome and requires either a dark room or an enclosed surface around the optics for operation. The use of ultraviolet detectors and light source could provide more accuracy. The use of optical systems does offer accuracy that may not be available with other systems. The above system could not sense the direction of change of Δx and should be used with increasing (or decreasing) force in one direction only.

When polarized white light passes through a photoelastic material, strain gradients cause a very slight shift in the speed of light because of the associated density gradient. Therefore, lines of constant strain may appear either dark or bright. A polariscope provides the polarized light, while the configuration to be studied is constructed of a photoelastic material (clear plastic). The photoelastic strain gage operates on this principle but does not require a polariscope. The incoming light is polarized by the gage, and the fringe patterns are residual strains built into the gage. The bibliography of reference 2 is a good source for additional reading in the area of strain measurement. The chapter on strain measurement is also recommended for more details on strain-measuring systems. The photoelastic strain gage has the advantage of immediately presenting a strain measurement in output form. It senses strain in a direction along its axis, or it can be made circular to sense the direction of maximum strain. The use of visual output requires an operator and a static condition. It is unlikely that this type of sensing system would be miniaturized to reduce the gage length (gage length is the unstressed length, x, of previous examples). The field of photoelasticity has been instrumental in the study of geometries that were previously unknown. However, the accuracy and sensitivity of these systems must be examined for each individual system.

The following systems can also be used to measure strain. They are mentioned here to remind the reader that the fundamental measurement is one of length and that an almost limitless quantity of sensors are available. If the anode of an electron tube is attached at one point on a specimen while the cathode is attached at another point, the number of electrons that will flow through the anode circuit is functionally dependent on the distance between these elements. Two capacitive plates can have variable separation by the same technique. Variable inductance, piezoelectric effect, piezoresistive effect, and simple electrical resistance can all be the basis of sensors for strain measurement. The number of high-energy particles passing through a sample is dependent on the lattice dimensions. The pressure of a liquid in a closed container or the thermal coefficient of expansion of a material can also be used to sense strain. However, the only other system to be discussed in detail is the simple electrical resistive element. The electrical-resistance strain gage is the most popular strain sensor ever developed.

THE ELECTRICAL-RESISTANCE STRAIN GAGE

The electrical resistance of a uniform cross-section wire is given by

$$R = \frac{\rho L}{A} \tag{7.7}$$

where R = resistance, ρ = electrical resistivity, L = wire length, and A = cross-sectional area. The fact that the electrical resistance of a wire changes with strain was first reported by Lord Kelvin in 1856. The total change in electrical resistance can be determined from Equation 7.7 if $A = (\text{const})(D^2)$, where D is a characteristic dimension of the cross-sectional area. [For the circular cross section the dimension D is diameter and the constant is $\pi/4$; for a rectangular cross section (x, y), $D = x$ and the constant is y/x.] Then

$$dR = \frac{L}{CD^2} d\rho + \frac{\rho}{CD^2} dL - \frac{2LR}{CD^3} dD$$

and

$$\frac{dR}{R} = \frac{d\rho}{\rho} + \frac{dL}{L} - \frac{2\,dD}{D} \tag{7.8}$$

Now dL/L = strain, and dividing both sides of Equation 7.8 by strain and rearranging gives

$$\frac{dR/R}{dL/L} = 1 - 2\frac{dD/D}{dL/L} + \frac{d\rho/\rho}{dL/L} \tag{7.9}$$

Using the definition of Poisson's ratio (Equation 7.6), Equation 7.9 becomes

$$\frac{1}{\epsilon}\frac{dR}{R} = 1 - 2\mu + \frac{1}{\epsilon}\frac{d\rho}{\rho} \qquad (7.10)$$

The electrical resistivity does not significantly change with strain for many materials. Using this fact, Equation 7.10 can be written

$$\frac{1}{\epsilon}\frac{dR}{R} = 1 - 2\mu = F \qquad (7.11)$$

where F is the gage factor for the electrical-resistance strain gage. If the material used is steel, $\mu \doteq -0.3$ and the gage factor is approximately 1.6. The gage factor varies with temperature, but it is essentially constant with respect to change in strain. Strain can be measured from the relationship

$$\epsilon = \frac{1}{F}\frac{dR}{R} \qquad (7.12)$$

The Wheatstone bridge circuit is normally used to measure the initial resistance and the change in resistance. However, a discussion of the sensor development will be presented before measurement techniques are covered.

One of the first wire-type strain sensors was simply a wire connected between two insulators attached to the specimen. Figure 7.15 shows this type of

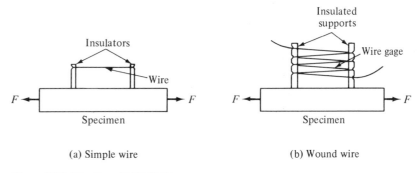

(a) Simple wire   (b) Wound wire

Figure 7.15 Wire-Type Strain Sensors

sensor, along with a modification that provided for increasing the wire length. These sensors were designed to measure tension but could be used for compressive strains by winding the initial wire in tension. Four problems associated with this type sensor are that the bonding of the insulated supports to the specimen is involved, the supports deform with increasing load, the value of initial wire resistance is very small, and the measurement of compressive strains is very limited. An improvement that helped to solve the first two of these problems was the SR-4 bonded-wire gage. This gage was also capable of measuring compressive strains. Figure 7.16 shows the

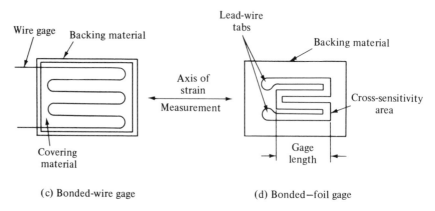

(c) Bonded-wire gage   (d) Bonded–foil gage

Figure 7.16 Bonded Strain Gages

essential elements of an SR-4 gage. It also presents the foil-type bonded strain gage which has become very popular because of the range of geometries, resistances, and gage factors available. In both gages the wire and foil are bonded to the backing material with a bonding strength greater than the strength of the gage material, if possible. The backing material is then bonded to the specimen with a bonding strength greater than this. Failure in the sensor is ensured to occur in an open circuit of the gage under these conditions.

The foil gages are made using printed-circuit techniques which allow flexibility in geometry and control of foil material and thickness. The variety of materials available for this use has made possible sensors with initial resistances of several hundred ohms. Control of manufacture and sample testing has reached the point where only gages with very precise initial resistance and gage factors are sold. Values of initial resistance, gage factor, and thermal strain coefficients are provided for each gage sold by the manufacturer. The accuracy limits on the values of these parameters are also given. Since the strain gage is functionally dependent on length, force, stress, strain, and temperature, it is a multipurpose sensor. The variety of applications of this sensor in the field of measurements justifies the extensive coverage that it will recieve here.

The foil-gage geometry presents two problems to the measurement of strain. Strain is to be sensed in one direction only when one gage is used. However, the gage material used at the cross-sensitivity area of Figure 7.16(b) senses strain in the direction perpendicular to the principal axis. The area of the foil is increased at this point, while the length of the gage in the cross-axis direction is minimized. These two effects tend to reduce the contribution of cross-axis strain to the total gage output. This does not entirely eliminate the cross-strain effects, and this fact should be remembered in the final analysis of data. A second geometry problem is associated with the gage length. In stress–strain analysis the strain at a geometric point is considered. The physical geometry of the strain gage makes it impossible to measure the strain at a point. Printed circuits make gage lengths from $\frac{1}{32}$ in. to 1 in. readily available. It is possible to manufacture much smaller gages by accepting larger errors in gage parameters, but geometrically the gage must have some physical dimensions. Therefore, the strain gage cannot measure the strain at a point. It does, in fact, measure the integrated effect of the strain present under the gage area. If the strain is constant over the area of the specimen where the gage is located, the gage measures the strain at any point under the gage. If the strain varies linearly along the axis of the strain gage, the gage measures the strain at the geometric center of the gage grid. In all other cases of strain variation, the gage indicates the strain at some point under the grid; but the exact point is not known.

The strain gage must be bonded to the test specimen for operation. The methods for bonding vary with specimen material, temperature range, and strain range. *The Strain Gage Primer* (reference 3) is recommended for instruction in gage mounting and use. After a strain gage has been bonded to a specimen, it cannot be removed and bonded to another sample. The act of bonding a gage to a surface does change the elastic properties of the surface. Thus, the first axiom of measurement is realized. This change in elastic properties is generally negligible for most applications, but it is of importance when the sample thickness is of the same order of magnitude as the gage and bonding-material thickness. Recommendations for surface preparation, bonding material, and weatherproofing are given by gage manufacturers.

The measurement of strain is affected by yielding, fatigue, and creep that occur between the various bonds and in the metal foil. Thermal effects are also present in all of these elements. Gages are constructed of backing material, bonding, and foils that are compatible with the ranges of stress for a given specimen material. When the elastic properties are not compatible with the specimen material, failure of the gage or decreased sensitivity results. Strict adherence to the manufacturer's specifications is recommended.

To obtain the state of stress at a point requires the determination of stress in three directions. Mohr's-circle techniques can be applied to obtain the information required. This means that three strain gages must be applied at a point to obtain the two-dimensional stress condition at this point. This problem is reduced by using strain-gage rosettes. The geometry of these rosettes varies, but the basic design incorporates the elements shown in Figure 7.17. One backing material has three strain gages bonded to it. These

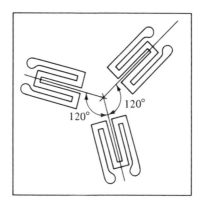 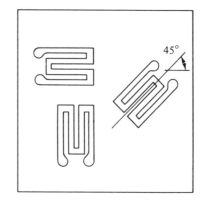

Figure 7.17 Strain Gage Rosettes

gages are usually oriented to measure strain at 120° intervals, although some gages are designed to have two directions mutually perpendicular with the third at 45°. In any case the rosette can determine the information required for measuring the state of stress at a point. When one or two simple gages are mounted on a specimen having no shear stress, the axis of the gage should be parallel to the normal stress (and lateral stress if two gages are used). Any angular error in the location of a gage axis is the same as an error of twice that angle when considered on the Mohr's circle. Alignment marks are placed on the basic gage to help eliminate any error in angular orientation. Mutually perpendicular alignment marks are also placed on the gage to locate the geometric center of the strain gage.

The lead-wire tabs are placed on the strain gage to facilitate the soldering of external lead wires to the gage. Low-temperature solder must be used in this operation to avoid destruction of the bond. The attachment of external leads is one of the most demanding operations in strain-gage mounting. The use of silver paint has been suggested, and other methods are becoming available. The use of a soldered joint is the most common practice. The joint is also subjected to yield, fatigue, and creep. The lead wire should be applied so that it will transmit zero force to the joint. The application of any force to the foil joint can easily fracture the foil and destroy the usefulness of the gage.

According to Equation 7.12, the electrical resistance varies linearly with strain. This is one major advantage of the electrical-resistance strain gage. Calibration could theoretically be accomplished by applying a known strain to the specimen on which the gage is mounted. Since the gage cannot be removed after mounting, the geometry of the specimen dictates the feasibility of this type of calibration. When the geometry is such that it is impossible to apply a known strain to the gage, electrical calibration is still possible. If the gage resistance is 100 Ω (a typical value), the addition of a 10^6 Ω resistance in parallel with the gage will cause a change in the electrical resistance:

$$\Delta R = 100 - \frac{10^8}{1,000,100} \doteq 100 - 99.99 = 0.01$$

Then, for a gage factor of 2,

$$\epsilon = \frac{1}{2}\left(\frac{0.01}{100}\right) = 0.00005 \frac{\text{in.}}{\text{in.}}$$

The use of a variety of parallel resistors can provide electrical calibration of the strain gage. The dynamic electrical calibration of a strain gage can be accomplished by incorporating an electronic switch which alternately connects and disconnects the parallel resistor into the strain-gage circuit. The dynamic response of an inductance–capacitance free circuit is limited only by the speed of electrons in the lead wires (approximately the speed of light).

The strain gage has excellent dynamic characteristics. It very closely approximates a zeroth-order system. This is one of its most valuable assets. However, the measurement circuitry incorporated in the overall measurement system and the recorder places some limitations on the total dynamics of the system. The use of shielded and grounded leads, electrical insulation over the gage, and an electrical noise-free environment can eliminate many of the dynamic response problems. The strain gage, the piezoelectric device, and the piezoresistive device all have excellent dynamic responses. All compare favorably with sensors limited by the speed of light (for a given medium) in dynamic response.

The intermediate circuitry normally used in measuring strain-gage output is the Wheatstone bridge. When only the dynamic signals are to be considered, the ballast circuit is also used. Since the Wheatstone bridge is nonlinear in operation, the analysis of the ballast circuit is simpler for dynamic signals. The limitation of the ballast circuit to dynamic situations is based on the accuracy requirement for this circuit. For the previous example, where $R = 100\ \Omega$ and $\Delta R = 0.01\ \Omega$, the ballast resistor would be $100\ \Omega$ for maximum sensitivity (see Chapter 6). If the input voltage is 10 V, the output voltage in the unstrained position will be 5 V. The output in the strained condition ($\Delta R = -0.01$) will be

$$e_o = 99.99\frac{10}{199.99} \doteq 4.9975$$

The requirement in measuring e_o is six significant figures. Using the Wheatstone circuit initially nulled ($R_1 = R_2 = R_3 = R_4 = 100\ \Omega$) the measurement of Δe_o will be

$$\Delta e_o = e_i \frac{\Delta R\ R}{4} = 10\frac{(1.0001)}{4} = 0.00025\text{ V}$$

This is only two significant figures. The cost of voltage-measuring devices dictates the use of the Wheatstone bridge for static strain measurement. The Wheatstone bridge circuit offers other advantages in the measurement of strain when more than one strain gage is placed in the bridge circuit.

Let us assume for the following discussion that all strain gages are identical and that strain varies linearly with $\Delta R/R$. Then the initial bridge is balanced by placing one strain gage in each arm of the bridge ($R_1 = R_2 = R_3 = R_4$). The Wheatstone bridge circuit is shown in Figure 7.18 for convenience. Initially, $e_o = 0$, since $R_1/R_2 = R_3/R_4$. The equation for the deflection bridge circuit (developed in Chapter 6) for initially equal arm

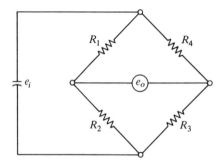

Figure 7.18 Wheatstone Bridge Circuit

resistances is (the $\Delta R/R$ terms in the denominator are frequently neglected)

$$\frac{\Delta e_o}{e_i} = \frac{(\Delta R_1/R_1) + (\Delta R_4/R_4) - (\Delta R_3/R_3) - (\Delta R_2/R_2)}{4 + 2(\Delta R_1/R_1) + (\Delta R_2/R_2) + (\Delta R_3/R_3) + (\Delta R_4/R_4)} \quad (7.13)$$

Now, if all four gages are initially at temperature T_1 and then at temperature T_2 the change in the resistance of each gage because of the change in temperature will be the same algebraically. Therefore, the net change in output voltage will be zero. The Wheatstone bridge provides temperature compensation for identical gages. The change in strain with change in temperature need not be considered when all four gages experience the same temperature change and all are arms of the bridge circuit. Temperature compensation is also provided automatically when two legs of the bridge are fixed while the other two are active legs, and the active legs are in adjacent arms (i.e., R_1, R_2; R_1, R_3; R_3, R_4; or R_2, R_4). This results for two active legs in adjacent arms experiencing the same temperature. The gages should also be attached to the same type of material. Temperature compensation using either two or four active legs (legs with strain gages in them) is a valuable asset of the Wheatstone bridge circuit, since it eliminates the temperature measurement and the computation that would otherwise be required.

Equation 7.13 can be applied to any system of strain-gage orientation on a specimen. The loading of the specimen can also be used in obtaining strain information. The examples of Figure 7.19 can be discussed for a variety of bridge connections. A brief discussion of each configuration will be given.

In part (a) axial strain varies linearly with position measured from the free end (stress $= mc/I$, $m = Fx$, $I =$ moment of inertia about the cross-sectional axis). Using Equation 7.13, with $R_1 = A$, $R_2 = B$, R_3 and R_4 fixed, $\Delta e_o \times 0$ since $\Delta R_1 = \Delta R_2$. Remember that $\epsilon_1 = (1/F)(\Delta R_1/R_1)$. If $R_1 = A$, $R_4 = B$, $\Delta e_o/e_i = (2F\epsilon_x)/[4 + (4)(F\epsilon_x)]$. If $R_1 = A$ and $R_4 = D$, $\Delta e_o/e_i = (F\epsilon_x + F\epsilon_y)/(4 + 2F\epsilon_x + 2F\epsilon_y)$. With $R_1 = A$ and $R_2 = C$, $\Delta e_o/e_i = [F\epsilon_x(1 + \mu)]/[4 + 2F\epsilon_x(1 + \mu)]$. The strain measured by gage C is negative Poisson's ratio times the strain measured by gage A.

In part (b) the strain is constant with respect to axial position. Therefore, $R_2 = A$, $R_3 = B$ gives $\Delta e_o/e_i = (2F\epsilon_x)/(4 + 2F\epsilon_x)$. With $R_1 = A$, $R_2 = C$, $\Delta e_o/e_i = [F\epsilon_x(1 + \mu)]/[4 + F\epsilon_x(1 + \mu)]$. With $R_1 = A$, $R_4 = C$, $e_o/e_i = [F\epsilon_x(1 - \mu)]/[4 + F\epsilon_x(1 - \mu)]$.

In part (c) each gage senses the maximum normal force, but one is in tension while the other is in compression. Therefore, $R_1 = A$. $R_2 = B$

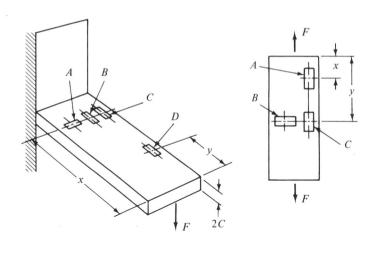

(a)   Cantilever beam

(b) Simple tension

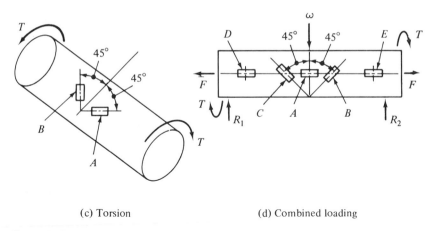

(c) Torsion

(d) Combined loading

Figure 7.19 Examples of Strain Gage Locations

gives $\Delta e_o/e_i = 2F\epsilon_n/4 + 2F\epsilon_n$. In part (d) with $R_1 = C$ and $R_2 = B$, the gages are on the neutral axis for ω. The effect of F would be positive for both gages (shearing stress $= \frac{1}{2}$ normal stress for effect of F only). The positive changes in resistance from the force F will algebraically cancel in legs R_1 and R_2. Therefore, the output with $R_1 = C$ and $R_2 = B$ will be only the effect of torque, T. If the gages are identical and linear, it is possible to consider the partial changes in resistance which result from each type of loading. A knowledge of Mohr's circle is very useful in the analysis of the output obtained from a Wheatstone bridge. See reference 2 for additional example.

Strain gages are frequently used to measure force and pressure in addition to strain. They are also used to measure torque, velocity, and acceleration in association with properly designed geometries. They are capable of measuring length change and temperature. Therefore, the strain gage could theoretically measure any dimension (since it can measure length, velocity, . . . , time, force, acceleration, . . . , mass, and temperature; all dimensions are composed of these). The strain gage is a versatile sensor with many capabilities.

PRESSURE

Pressure, like stress, is defined to be a force applied over a given area. Stress is associated primarily with solid mechanics, while pressure is related to fluid mechanics. Since they are essentially the same physical quantity, some terms from each field are frequently used in the other field. The primary difference between a fluid and a solid is that any shearing stress will cause continuous deformation of the fluid, while the solid will experience a fixed deformation under a shear load. Fluids generally cannot support a tensile force (some experiments have been devised to measure the tensile force that can be applied to a liquid before change of phase is initiated, but this is not the ordinary situation). Pressure is, therefore, essentially a compressive force and is zero when zero constraining force exists. For a contained gas, pressure is a measure of the change of momentum of the gas molecules when they collide with the container wall (kinetic theory). In static liquids pressure is a measure of the force required to contain the liquid plus the gravitational attraction of the liquid.

The methods employed in measuring pressure have introduced a unique terminology into the pressure-measurement field. We shall soon see that many pressure-measuring systems actually detect only a differential pressure between the desired signal and the environment. Usually, the environment is the local atmospheric pressure (not that corrected to sea level, as provided by weather stations). When a pressure-measuring system measures differential pressure referenced to the local atmospheric pressure, the output reading is called gage pressure. When a pressure-measuring system does not depend on the environmental conditions, the output is usually referenced to the absolute zero on the pressure scale. This pressure is called the absolute pressure. Figure 7.20 shows the two reference pressure scales that are

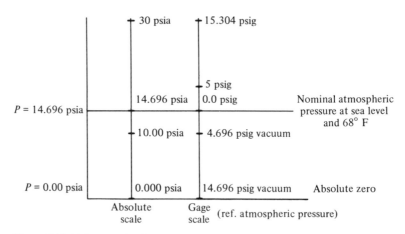

Figure 7.20 References for Pressure Measurement

used in practice. The absolute pressure 14.696 psia is called the standard atmosphere. A standard atmosphere will also support a column of mercury 760 mm high (i.e., 14.696 psia = 760 mm Hg at 68°F). Three other commonly used pressure terms have the following definitions: 1 mm = 1333.22 microbars = 1333,22 dynes/cm²; 1 mm Hg at 68°F = 1 torr; 1 mm Hg at 68°F = 1000 μ. When a pressure measurement is referenced to the local atmospheric pressure, the dimensions on this reading are usually given in pounds per square inch gage (psig). If the pressure measured is less than atmospheric pressure, the reading is usually given in psig vacuum. Either the word vacuum or a negative differential pressure must be given to clarify the fact that the measured pressure is below atmospheric pressure.

The pressures in the range 0.02–12,000 psia are called normal pressures for this book. Lower pressures are presented in the discussion on vacuum, while higher pressures are usually considered when using solid mechanics sensors. The basic standards for normal pressures are the water and mercury manometers and the dead-weight tester. These systems will be discussed in some detail to provide a foundation for calibrating and analysing other measurement systems in this range. These systems are presented for static pressure measurement.

Liquid monometers are available in various forms. Some of the possible confiugrations are shown in Figure 7.21. In all cases the measurements that must be made include the areas of the tubes and the distances between the liquid levels. The basic analysis of a manometer is a simple force balance. For the U-tube manometer the force balance uses the fact that the pressure

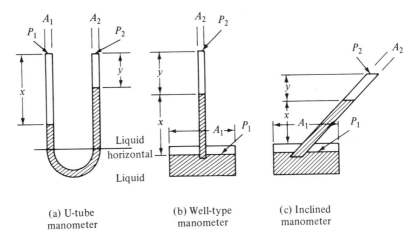

(a) U-tube
manometer

(b) Well-type
manometer

(c) Inclined
manometer

Figure 7.21 Liquid Manometers

at any horizontal line drawn through both legs of the manometer is a constant in the liquid. Then the force balance is

$$P_1 A_1 + x A_1 \text{(density of media in the region)}$$
$$= P_2 A_2 + y A_2 \text{(density media)} + (x - y)(A_2)\text{(density liquid)}$$

The density in all cases must be multiplied by the local acceleration of gravity and divided by the gravitational constant, g_c, to change from mass density to weight per unit volume. Five parameters must be considered. The measurement of length and area have already been discussed. The extraneous inputs affecting these measurements also cause extraneous inputs in the measurement of pressure. Temperature is a major error source. Precision-area tubes are desirable. Density varies with impurities, temperature, and pressure. Impurities should be removed, and density of the liquid varies very slightly with pressure. However, temperature corrections should be made. Any error in determining the local acceleration due to gravity will introduce an error in the analysis of the manometer. Typically, gravitational data based on local elevation and latitude are used for this correction. Local acceleration is also considered in Section 7.5. The liquid interface lines in Figure 7.21 are horizontal. However, the effect of surface tension at most interfaces between liquids and solids causes the interface to be curved. The resulting curved interface is called a meniscus and is shown in Figure 7.22. Usually, mercury is measured at the top of the meniscus, while water is measured at the bottom. Knowing the tube diameter and the temperature, the appropriate corrections are given in reference 5. When

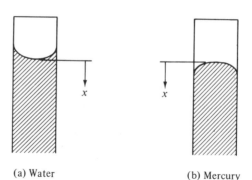

(a) Water   (b) Mercury   **Figure 7.22** Liquid Meniscus

a U-tube manometer is used with both legs having the same diameter, the meniscus effect cancels out. Finally, one pressure must be known before another pressure can be determined. This fact leads to the need for a reference pressure. One solution to the problem is to completely evacuate one side of the U-tube manometer so that $P_1 = 0.0$ psia. Operating in this mode, the measurement system is called a barometer.

The mercury barometer is an essential part of any laboratory. If one side of a U-tube is sealed (or the single leg of the well-type manometer), filled with mercury by tilting, and then turned upright, the pressure above the mercury in that leg is zero. The vapor pressure of mercury is low enough to be neglected. (The vapor pressure of water and the column height required virtually eliminates water for use as the barometer fluid.) Many barometers are designed for use at a given location by precision boring of the leg area and the well area to correct for the local acceleration of gravity (reference 4). Proper selection of the method of length measurement can provide readings with an accuracy of ± 0.001 in. These methods incorporate micrometer, vernier scales, microscopes, and optical methods. If mercury is the fluid, the pressure is accurate to approximately ± 0.0005 psia. Other manometer fluids, such as oil and alcohol, are available. However, mercury and water are the standards.

Figure 7.23 shows the U-tube manometer with three fluids under the same differential pressure. The need for a fluid with density between mercury and water is obvious. Accuracy capabilities and range of operation for water and mercury are 0.0–30 in. of water ± 0.004 in. and 0.0–100 in. of mercury ± 0.025 in. When a manometer is inclined, the length measurement is larger; but the meniscus is formed over a larger effective diameter. The overall accuracy is improved if the method of measuring length is a major source of error. The measurement of length using a scale placed between the legs of a U-tube manometer is subject to many errors. For this system

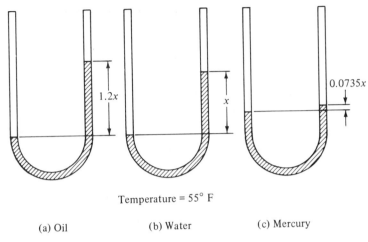

Temperature = 55° F

(a) Oil   (b) Water   (c) Mercury

Figure 7.23 Manometer Reading for Three Liquids with One Pressure

the error could easily be ± 0.25 in. Three possible methods of measuring length are shown in Figure 7.24. In part (a) the pointer will change the curvature of the interface on contact. This method is commonly used with mercury barometers to adjust the scale with the mercury level in the well. Part (b) shows a horizontal bar with back light. When no light passes the top of the meniscus, the reading is taken. In part (c) the liquid level is determined from the geometry of the optical system. The effect of interface

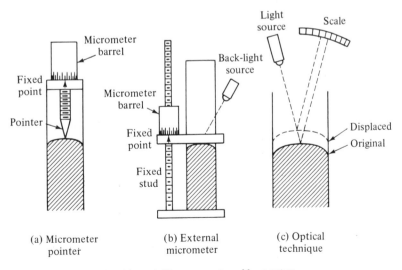

(a) Micrometer   (b) External   (c) Optical
    pointer   micrometer   technique

Figure 7.24 Methods of Length Measurement on Manometers

curvature can be eliminated by always adjusting the light source and measuring on the light-source support.

A second laboratory standard for static pressure calibration is the deadweight tester. The accuracy capabilities of the dead-weight tester is given for various ranges: 0.3–50 psia ±0.01 psi, 50–500 psia ±0.125 psi, 500–2417 psia ±0.25 psi, and 2417–12141 psia ±0.25 to ±6.5 psi (reference 4). The dead-weight tester is also a force-balancing system. The basic elements are shown in Figure 7.25. A known weight is supported by the liquid in

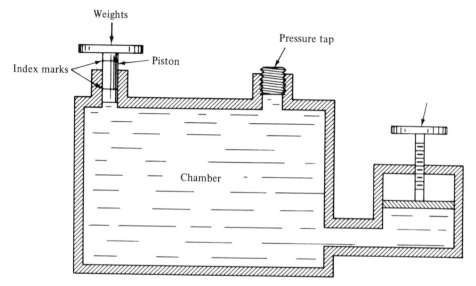

Figure 7.25 Elements of a Dead-Weight Tester

the chamber. The area of the piston is carefully measured, and the pressure applied to the liquid is the weight divided by this area. The index marks on the piston are adjusted so that one index mark shows while the other is inside the cylinder. The pressure tap allows the measurement system to be calibrated across to the standard pressure chamber.

Six possible error sources for this system are to be mentioned here. The area of the piston–cylinder arrangement must be measured accurately. Temperature effects on expansion must be considered, and the local acceleration of gravity must be used to find the weight. These errors are analyzed by measurements discussed previously and by quality workmanship. The piston experiences a buoyancy effect of the weight of the fluid displaced by the piston. The index marks help to maintain the value of this effect constant. The static friction of the liquid in the chamber causes the piston to be partially supported by the shear stress in the fluid between the piston and the cylinder. This effect would be zero after equilibrium has been reached, but the piston is usually given an angular rotation to eliminate this partial support. Any machining marks or scratches on the piston cause some error because of this rotation. Therefore, the piston should be rotated in both directions to determine if any piston-surface defects exist. Finally, if the load is not applied vertically, the piston could be partially supported by the sides of the cylinder. The results of eccentric loading can be eliminated by a quality tester and proper application of weights on the piston.

The dead-weight tester can be calibrated in the lowest pressure range using a mercury manometer. Pressures below atmospheric are obtainable with a vacuum environment around the dead-weight tester. The effect of buoyancy, local gravity, and loading are measured in the calibration process. The dead-weight tester is then ready for use in calibrating other pressure-measurement systems. A brief discussion of measurement systems for normal pressures follows. The operation and calibration of these systems is principally for the measurement of static pressure. Dynamic calibration is discussed in Appendix III.

SENSORS FOR THE NORMAL PRESSURE RANGE

The Bourdon tube, the diaphragm gage, the piezoelectric crystal, and variable-electrical-properties gages are all commonly used to measure pressure in the normal pressure range. These systems will be introduced here along with the various capabilities of each.

The Bourdon tube is shown in Figure 7.26 with a linkage system to convert motion of the sealed end of the tube into rotational motion of a spring-loaded pointer. The principle of operation is the fact that an oval cross-section curved tube will attempt to straighten when the pressure inside the tube is greater than the pressure outside. (It will tend to be more curved when the pressure inside is smaller.) Therefore, the Bourdon tube measures gage pressure (i.e., psig) and the pressure surrounding the tube must be known. Although there is a functional relationship between the pressure inside the tube and the motion of the end of the tube for given environmental pressure [$P = f$(position)], the mathematical model for this relationship is not easily established. However, this fact in no way affects the operation of the system once static calibration (using either a manometer or dead-weight tester) has been performed. Extraneous inputs include linkage friction, temperature, any obstruction of the inside of the tube, and

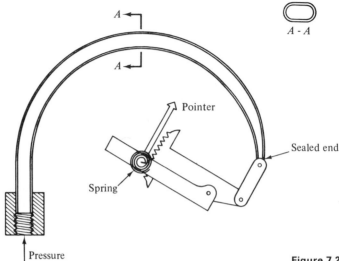

A - A

Pointer

Sealed end

Spring

Pressure

Figure 7.26 Bourdon Tube

lost motion in the linkage. Calibration can be used to analyze most of these input signals. Calibration with both increasing and decreasing pressures is important, and a tap on the gage can help to reduce friction effects. If the gage cannot be tapped in actual operation, it should not be subjected to this condition in calibration. The accuracy of the Bourdon tube can be in the range of ± 0.1 to ± 0.5 per cent of full-scale reading. The nonlinearity of the gage is partially corrected by linkage design. The range of a Bourdon tube is 0.0–100,000 psig. The Bourdon gage is simple to use, relatively inexpensive, and contains both intermediate modifying systems and read-out device. It is basically a static pressure-measuring device.

The diaphragm and bellows gages transduce a pressure signal into a strain and a displacement. In Figure 7.27 the two sensors are shown. The output of either can be used in conjuction with many types of sensors for strain and length to provide a complete system of pressure measurement. The type of additional sensor used with these basic sensors determines the capabilities of the sensors to operate for static and dynamic pressure measurement. Strain gages and variable-electrical-property gages may be used in dynamic operation. In all dynamic pressure measurement the geometry of the basic sensor and the connecting parts introduce extraneous inputs into the dynamic signal. A basic study of fluid dynamics is essential to understanding the effect of geometry on the dynamic signal, but dynamic calibration is the method ultimately used to establish consistency between input and output signals. Both basic sensors measure differential pressure, which is usually gage pressure. The accuracy and range of these systems are

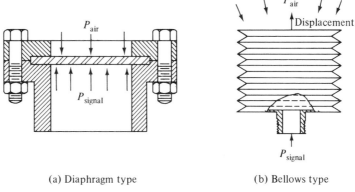

| (a) Diaphragm type | (b) Bellows type |

Figure 7.27 Diaphragm and Billows Type Sensors

essentially the same as those of the Bourdon gage. The intermediate circuitry and read-out equipment depend on the type of secondary sensor employed. Temperature and stress geometries are extraneous inputs. If the diaphragm is nonuniform or nonhomogeneous, the stress will have nonlinearities in addition to those present in the original geometry. All extraneous inputs associated with the secondary sensors will also be present in this system. Many of these inputs have been mentioned where the sensors were originally presented.

The variable-electrical-property sensors include slide-wire resistors, variable capacitors, variable inductors, and mutual inductances. The slide-wire resistor and linear variable-differential transformer were discussed earlier. The other two devices will be discussed in the last section of this chapter. Other types of secondary sensors are available. The entire field of measurements can be used to develop a system for the measurement of any parameter. System designs are limited only to the imagination of the developer. When a functional relationship exists between the parameter to be measured and a measurable parameter, a measurement system can be constructed to measure the desired parameter. It is not necessary to know completely the functional relationship that exists for a system to provide satisfactory operation. The calibration of the system establishes the functional relationship required.

The fused-quartz Bourdon gage was developed for precision measurement of pressures in the range 0.1 mm to 4 m of mercury. The fused quartz is wound helically, so that a pressure change inside the Bourdon-shaped quartz tube causes an angular rotation. The method of output measurement is very similar to the D'Arsonval-meter movement in that the output sensor may be a light source with a mirror mounted on the quartz element. The quartz is relatively free from hysteresis, fatigue, and creep, and has relatively constant thermal-expansion coefficients and thermoelectric modulus. These properties help to reduce the effect of extraneous inputs and to result in accuracies equivalent to the manometer and dead-weight tester.

The piezoelectric pressure transducer has many very important advantages when used for dynamic pressure measurement. Several crystals experience a reorientation of the electrons in the lattice structure when the structure is mechanically stressed. Some of the piezoelectric materials displaying the electrostatic shifting of electrons with stress occur naturally (quartz and tourmaline), while many others have been developed in an artificial environment (Rochelle salt, barium titanate, lithium sulfate, and others). The sensor is supported in a variety of ways. One method is shown in Figure 7.28. Here the electrons shift to one of the collector plates, depending

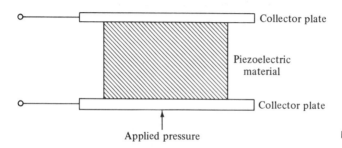

Figure 7.28 Piezoelectric Sensor

on whether tension or compression is applied to the piezoelectric element. The piezoelelctric material must be cut properly along crystallographic axes (for crystals), and ceramics must be properly heat treated and polarized. During a dynamic application of stress, electrons migrate to one collector plate. Once the static stressed state has been attained, the electrons tend to move back to their original position in the lattice structure. Therefore, design of collector plates and intermediate circuitry has been a very important aspect in using piezoelectric sensors. The sensors respond dynamically in the range 10–50,000 Hz. At lower frequencies the electrons tend to leak back to their initial positions, while the higher-frequency response is limited by the rate at which the electrons move in the lattice structure as a result of applied stress.

Extraneous inputs are temperature, acceleration, vibration, humidity, electrical noise, and extraneous pressure waves. The rate of change of temperature also introduces an extraneous signal. Since the signal output has a transducer impedance of 1 MΩ, a charge amplifier and voltage amplifier are required to obtain impedance matching of sensor to output device. The range is limited by material strength and thermal ranges (-400 to $+500°F$). The accuracy can equal ±0.01 psia for limited pressure ranges. The strain gage on a diaphragm compares very favorably with the piezoelectric sensor.

SENSORS FOR THE HIGH-PRESURE RANGE

Several of the systems already mentioned can be suitably adapted to accommodate higher pressures. The dead-weight tester has already been discussed in detail. The use of a diaphragm was necessary at low pressures to increase the deflection and resulting strain at these pressures. For high pressures the actual container can be used for the diaphragm. The sensitivity is reduced, but the principle of operation is the same. Any system that can accept a high pressure over a known area can sense the signal using a simple force-measuring system. The proving ring is one possible sensor for high pressures. The circular shape of the proving ring and several properly located strain gages provide a flexible sensor with relatively high output. Some Bourdon-tube gages are available for high pressures.

One slightly different type of sensor is shown in Figure 7.29. The electrical resistance will change because of the bulk compression that results

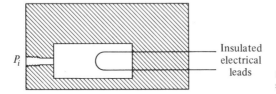

P_i — Insulated electrical leads

Figure 7.29 Electrical Resistance Pressure Gage

from exposure to high pressures. If the fluid in the chamber is nonconducting (kerosene can be used), the total resistance change is the same as the analysis for strain gages, where stress is replaced by pressure. The sensitivity for most high-pressure sensors is limited by the design. The range is governed by the material strength. Extraneous inputs are temperature and electrical noise, but the high pressures present help reduce the problems of extraneous pressure signals. Yield, fatigue, and strain concentrations must be considered.

SENSORS FOR THE LOW-PRESSURE RANGE

An absolute pressure lower than 1 mm of mercury is difficult to measure using manometers because of change of phase and sensitivity problems. The variety of vacuum-technology applications has created a very broad spectrum of sensors with a variety of operating principles. Here an attempt will be made to present the sensors, their principles of operation, and their ranges. The physical meaning of pressure may be a serious question for pressures below 10^{-3} torr (1 torr $= 1$ mm Hg; 10^{-3} torr $= 1\ \mu$). However, the answer to questions of this type will not be attempted here. The problem is the sensor, how it operates, and how it can be calibrated.

The primary standard for the measurement of low-pressure static signals is the McLeod gage. It is shown in Figure 7.30. The operating principle is

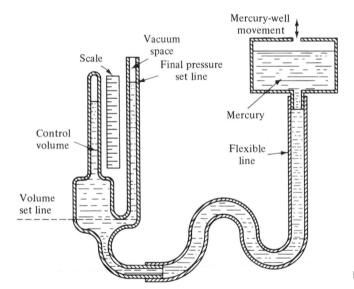

Figure 7.30 McLeod Vacuum Gage

the ideal gas law for isothermal constant-mass operation, $P_1 V_1 = P_2 V_2$. When the mercury well is lowered until the mercury level is below the volume set line, the known control volume is exposed to the vacuum space being measured. The mercury well is raised to the volume set line, isolating the unknown pressure of the vacuum space in the known control volume. This fixed mass of molecules is compressed to a final known pressure by lifting the mercury well until it reaches the final pressure set line. The final volume of the mass of molecules in the control volume is measured on the scale. Knowing the initial volume, V_1, the final pressure, P_2, and measuring the final volume, V_2, determines the initial pressure, P_1. The scale is an analog scale which reads the initial pressure instead of the final volume.

The McLeod gage is the primary standard for the range 10–10^{-6} mm Hg. It is used to calibrate practically all vacuum gages over some range of their operation (reference 6). There will be some pressure gradient in the internal geometry of the McLeod gage. This means the pressure reading will be higher than that actually in the vacuum space. This error can be greatly reduced by leaving the gage attached to the vacuum space for a long period of time before a pressure measurement is made. The pressure should not be measured when the vacuum space has a rapidly decreasing pressure in progress. Here the trapped pressure in the gage would be much higher than the space pressure at a later time. The trapped pressure will frequently blow mercury back into the vacuum space when the mercury well is lowered near the volume set line.

Extraneous inputs are any conditions that would change gage volume, scale position, or pressure levels. Some simple types are temperature, external pressure, and contamination in the mercury. Scales are usually movable and may require calibration at higher pressures using an inclined mercury manometer. Calibration of other instruments uses a series of differential pressure chambers with various pressures and variable leak rates between them (reference 6). At very low pressures, metals give up absorbed gases (outgassing) to cause some extraneous inputs. A more complete discussion of calibration in the 10^{-4}–10^{-10} torr range is given in reference 7.

A second type of sensor operates on the principle that the thermal conductivity of the gases in a space is strongly dependent on the geometry and the pressure in the range of 1–10^{-3} mm Hg. If an electrical wire is heated at a known rate by a control power source, the temperature of the wire will reflect the rate of heat transfer to the gas. When a thermocouple is attached to the wire (or is the heated filament), the gage is called a thermocouple gage. If the wire temperature is measured using a thermistor, the pressure gage is called a thermistor gage. And when the wire temperature is measured by the change in wire resistance with temperature, the gage is called a Pirani gage. All three types vary only in the method of measuring the filament-wire temperature. All require relatively complex power sources and modifying circuitry for measuring temperature. These gages are for pressures in the transition region between continuum flow and free molecular flow. This region is called slip flow. In both free molecular flow and in continuum flow the thermal conductivity of the gases is almost constant. In slip flow there is a relatively large change in thermal conductivity with pressure.

The response time of these gages is in the range 2–30 s. The cost is in the range $100–200. Most of this cost is associated with the power source and temperature-measuring equipment. The type of gas in the space can change the static output signal. This could be an extraneous input to the sensor. Temperature is also an extraneous input. Figure 7.31 shows the basic elements of a thermocouple gage. The wire is heated using a constant-energy

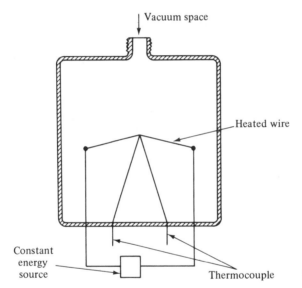

Figure 7.31 Thermocouple Gage

source, and the variation in temperature depends on the rate at which energy can be removed from the wire (primarily by interaction with the gas).

The ionization gage shown in Figure 7.32 operates on the principle that high-energy electrons will ionize the gas particles in a space. The cathode may be heated to help drive off electrons, or it may be an unheated (cold) cathode. The potential difference between plate and cathode must be high enough to give electrons enough energy to ionize the gas in the space. For a given geometry, the number of ions formed will be proportional to the number of gas molecules in the space. This then is a measure of the pressure. The number of ions collected will be a measure of the pressure. A wide variety of collection methods, grid properties, and ion-collector properties are possible (references 4 and 8).

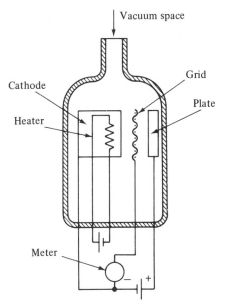

Figure 7.32 Ionization Gage

The general range of pressures for the ionization gage is 10^{-2} mm Hg to below 10^{-10} mm Hg. When operated at higher pressures, the number of ionizations can increase the current flow to the point where the collector elements are damaged. The type of gas present influences the number of ionizations that can occur. Gas particles are adsorbed by the metal elements when exposed to normal pressures. Usually, degassers in the form of heater elements are placed in the gage to drive off adsorbed molecules.

The ionization gage also collects gas molecules when they are attracted to the cathode. This is called ionic pumping. The gage operates to decrease the pressure in the system. This gage is the principle gage for pressures below 10^{-7} torr. Temperature, type of gas in the system, ionic pumping, pressure gradients, and gas adsorption are all extraneous inputs. The gage control unit, including various power sources and current meter, is relatively expensive (approximately $300–$500). The sensor and control unit can easily be damaged when used above 10^{-3} torr. Some units incorporate automatic shutdown when exposure to high pressures occurs.

When ionization is produced by alpha particles from a radioactive source, the gage is called an "alphatron." The method of operation is the same as other ionization gages except that the cathode is not required. The radiation source is constructed so that a constant rate of production of alpha particles is possible. The range of this gage is $10^{3}–10^{-13}$ torr. Here the problem of filament burnout and ionic pumping are essentially eliminated. The use of a radioactive element involves some precautions in the space where it is used.

Measurement in the low-pressure range has become increasingly important with space applications. Many sources are available for extending the present discussion. The references listed in this section are a beginning. Other texts on vacuum technology are available and should be studied for a more comprehensive coverage.

TORQUE

The last force- and length-type system to be mentioned is torque, which is the moment of a force about a given point. Mathematically, the summation of all moments about a point is equal to the summations of all forces times their lever arms,

$$\sum M = \sum F \times r = \text{torque}$$

A torque system is capable of measuring force, length, and the angle between the line of action of the force and a line through the point where moment is to be measured. Many possible systems have already been discussed.

For any measurement system involving the basic dimensional quantities, the extraneous inputs are the same as those discussed for each basic dimensional quantity. Therefore, torque is influenced by an extraneous input for force and length. One must be careful to recognize this fact and to apply this knowledge of any physical measurement system.

The beam balance is actually a torque-balance system. The cantilever beam is also a torque-measuring device. Both essentially measure static torque. The torque wrench of Figure 7.33 also measures static torque. This is one design of many possible. The pointer is attached to the bolt drive and the scale is attached to the lever arm. An applied force deflects the lever arm and scale relative to the bolt drive and pointer. Static calibration

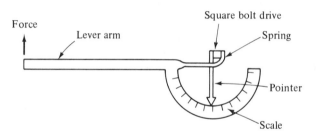

Figure 7.33 Torque Wrench

can be accomplished by measuring the force and geometry of the wrench.

Most of the systems for measuring energy (in foot-pounds force) are presented in Chapter 8, but the measurement of dynamic torque is also used for this purpose with rotating systems. Torque measurement and angular-rotation measurement can be used to determine the power (in foot-pounds force per second) of rotating systems.

$$\text{power} = (\text{torque})(\text{rotational speed})$$

This is usually a braking power when the torque results from loading the output shaft of a motor. This type of torque meter is called a dynamometer. Two mechanical dynamometer systems are shown in Figure 7.34. Both use

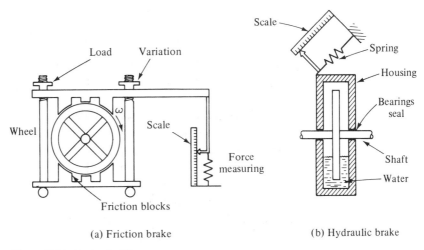

(a) Friction brake   (b) Hydraulic brake

Figure 7.34 Mechanical Dynamometer

mechanical methods for loading the output shaft and measure the resultant force at a known radial distance from the center of the shaft. Part (a) shows a prony brake, while part (b) shows a water brake. In part (a) the load is varied by tightening the wing nuts, and the water level in part (b) controls the resultant friction forces.

Electrical dynamometers are used in most present measurement systems. They inherently offer more reliable loading control, particularly at low speeds. Two types are most common. The first dissipates the energy internally in the measurement system (eddy-current dynamometers). The second type converts the rotational energy into electrical energy using either an ac or a dc generator. The electrical energy is dissipated externally to the system.

In all types the mechanical friction resulting from the shaft mounting method introduces some error in power measurement. The measurement of torque, voltage, and current are also subject to error. A complete thermodynamic analysis of the energy modes must be accomplished to provide accurate data for the measurement system.

Four parameters that are fundamentally important for many real engineering problems are linear velocity, linear acceleration, angular velocity, and angular acceleration. Any problem involving motion requires a knowledge of at least two of these parameters. These parameters are all vector quantities; therefore, they must be determined with both magnitude and direction. Force and pressure also require the evaluation of direction. However, the discussion here will be concerned only with the magnitude, since direction is inherently covered in directed-length measurements.

All four of the parameters to be discussed in this section are, mathematically, derivative-type quantities. That is, they must be evaluated instantaneously in order to reflect the truly defined parameter. All physical systems are limited in response by the speed of light. For many typical engineering problems this is actually no limit on the measurement process, because the "normal" velocities encountered are several orders of magnitude smaller than the speed of light. The speed of sound (pressure propagations in a given medium) may also be a limiting factor in some problems.

Several different types of sensors will be introduced for each parameter. It would be impossible to completely cover the field of sensors because of the many possibilities that are available. It is desirable to cover the most commonly used systems along with the more sophisticated.

LINEAR VELOCITY

The magnitude of velocity is called the speed. It is the derivative of length with respect to time. One of the most familiar speed indicators is the speedometer in an automobile. This system actually measures angular velocity and uses kinematic relationships to establish the linear velocity. Therefore, it will be mentioned in the section on angular velocity.

When position is changing in one direction, several methods of measuring velocity are easily visualized. High-speed photography of the moving mass with photographing speeds up to 1000 times the speed of the mass can be used to evaluate an average velocity by measuring the displacement from one picture frame to another and knowing the time required for one frame to be recorded. Camera speeds above 100,000 frames/s are possible. Closely spaced photoelectric cells and oscillograph records can be used to measure the passage of a mass. Radar uses the reflection of electromagnetic waves by a mass to measure the speed. Here the time required for a wave to travel from a source to the moving mass and back to a receiver determines the position of the mass as a function of time (the speed of light is the limit). Chopped visible light can be reflected by the object to accomplish the same purpose. Figure 7.35 shows the operation of these types of velocity-measuring systems. The motion picture can measure the displacement for 1/100,000 s. The photoelectric cell can measure the time required to travel each distance, x_i; and the electromagnetic wave traveling at the speed of light can be used to measure the displacement as a function of the chopping frequency.

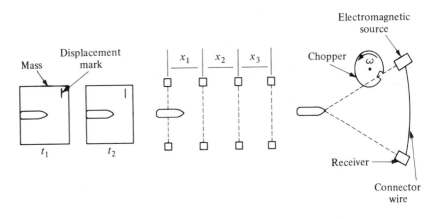

(a) Motion-picture camera   (b) Photoelectric cells   (c) Electromagnetic waves

Figure 7.35 Velocity Measuring System

If position as a function of time can be measured, differentiation can be performed to determine velocity. It is also possible to measure acceleration as a function of time and integrate to determine velocity. Usually, integration gives a better time-average value of velocity. Figure 7.36 shows another method of measuring linear velocity in terms of angular velocity. Angular velocity is an easier measurement problem, because the sensors can be located in a fixed reference frame. For a limited range of displacement, the slider motion on an electrical-resistance wire can be used to measure velocity.

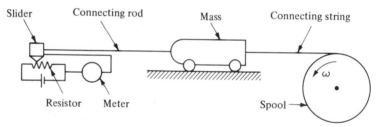

Figure 7.36 Angular-Linear Velocity System

Signals from angular-velocity measurement systems and linear-acceleration measurement systems can be used to obtain linear velocity. Many other physical laws involving velocity are available for developing measurement systems. A coil of wire moving through a magnetic field will give an output current depending on the velocity of the coil. Once the principle of operation of a sensor is known, calibration and elimination of extraneous inputs can be accomplished by application of previous methods for time and length.

Another interesting method of measuring velocity occurs in nature in the form of electrical storms. When lightning strikes, the time required to observe the occurrence is the time required for light to travel from the source to an observer. The pressure waves (thunder), on the other hand, can be sensed by the observer in the time required for sound to travel the same distance. Since light travels approximately 1 million times faster than sound in air, the distance to the source is approximately equal to the velocity of sound times the time between when the observer sees the light and when he hears the thunder. If a flash-sounding producing source is placed on a mass, photoelectric cell and a microphone can be used to measure the change in position with respect to time. The time is determined by the frequency of flashing of the source.

ANGULAR VELOCITY

Several methods of measuring angular velocity are shown in Figure 7.37. The principles of operation vary over a variety of possibilities. The mechanical counters of part (a) may have an associated stop watch-and-clutch assembly to engage the counter for a fixed time interval. This device is capable of measuring only average velocity. The photoelectric detector of part (b) will receive a signal each time the disc on the shaft exposes a white mark to the light source. This output signal can be used with either an electronic counter or an oscilloscope to obtain the time between pulses. Both the

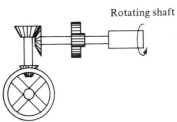

(a) Mechanical counters

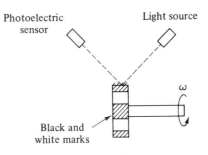

(b) Reflected light

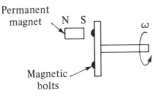

(c) Magnetic pick-up

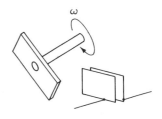

(d) Capacitive pick-up

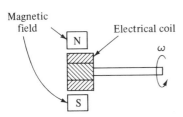

(e) ac or dc generator

(f) Centrifugal force

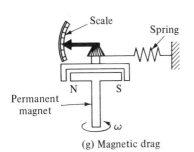

(g) Magnetic drag

(h) Stroboscope

Figure 7.37 Angular Velocity Sensors

magnetic pick-up and the capacitive pick-up [part (c) and (d)] experience a change in electrical properties with shaft position. Modifying the signal and use of the electronic counter or oscilloscope would provide a measure of angular velocity.

Both ac and dc generators [part (e)] give an output signal that is proportional to the input rotational speed. The rate at which either an electrical coil or a slotted disc of magnetic material passes through the external electrical field or magnetic field determines the generated signal. For part (f) the centrifugal force on the concentrated masses increases with rotational speed. This force is resisted by the spring, and the resultant static displacement (for constant angular velocity) is transmitted by a lever to a pointer–scale. This is basically a static angular-velocity measuring device.

Most automobile speedometers use the magnetic drag-type sensor of part (g). The rotating magnetic field produces rotation of a magnetic material disc due to eddy currents. The disc is spring loaded to result in a static balance of forces. This particular rotational speed is proportional to the rotational speed of the automobile tires. Therefore, a fixed tire diameter and a no-slip condition between tire and road can be used to measure linear velocity. All other methods of measuring angular velocity could also be used for this purpose, but the magnetic-drag method is relatively inexpensive and precise enough for automobile speed.

The timing light used to adjust internal engine performance is a form of stroboscope [part (h)]. When the light is flashing at the same frequency as the shaft rotation, the shaft appears to stand still. This is also possible when the light flashes at any integer multiple frequency or at any fractional frequency. With one dot on the shaft, two dots will appear diametrically opposite when the light flashes at $\frac{1}{2}, \frac{1}{4}, \frac{1}{8}, \ldots$ the rotational speed of the shaft. One dot appears fixed when the light frequency is $1, 2, 3, \ldots$ times the shaft rotational speed. An adjustment of the flashing frequency can be used to obtain the one-to-one ratio.

One other method of measuring rotational speed incorporates a circular-resistance slide wire. An electrical contact slider is fixed to the rotating shaft; the resistance between one end of the slide wire and the slider varies linearly with the position. This device is frequently used with low-rotational speed systems. The wear resulting from contact is a problem at high speeds.

LINEAR ACCELERATION

When the parameter force was introduced, it was mentioned that the local acceleration of gravity is an essential element in the definition of force.

Therefore, every measurement that involves force (this also includes the absolute ampere and volt) inherently depends on the value of local acceleration due to gravity. This measurement is not to be confused with the dimensional constant, $1 \, lb_f = 32.17405 \, \text{ft-lb}_m/\text{s}^2$. The dimensional constant is arbitrarily defined and internationally accepted. The fact that it is, numerically, approximately equal to the local acceleration at sea level on the forty-fifth parallel of latitude in no way alters the conceptual meaning. The dimensional constant does not change with either time or space. On the other hand, the local acceleration of gravity varies with both parameters to some extent.

The first measurement of absolute gravitational acceleration was made at Potsdam, Germany. This used reversible pendulums (reference 9). A later method used freely falling masses in a vacuum to determine local acceleration. Gas lasers and interferometer techniques will apparently replace the above methods. In the United States the Coast and Geodetic Survey records and maintains local gravity in much the same form as it does local elevations. This agency can provide a laboratory with local acceleration with an accuracy of $\pm 0.00004 \, \text{m/s}^2$ when provided latitude and longitude to $\pm 0.1'$ of arc and elevation within $\pm 5 \, \text{ft}$ (reference 9). If a more precise determination is required, local experiments must be performed. This is expensive and involves a considerable amount of work at the present time. However, mobile laser systems are available and most standard laboratories have precision gravity determinations.

Two methods of calibrating acceleration-measuring devices require local acceleration. These are the simple static turnover tests and the rotation of a vertical disc at a constant speed (reference 10). The rotating-disc method is discussed in detail in reference 10. Two other methods possible in calibration are the mechanical (or electrical) shaker table and the impact drop test. The rotating disc and the shaker table can be used to calibrate time-varying acceleration, while the turnover and drop test are for single-value or static acceleration.

Any mass moving under the influence of an applied force is theoretically capable of measuring acceleration. Some examples already discussed are the cup of water, the cantilever beam, the spring–mass, and a strain gage on a diaphragm. These systems are all basically second-order systems, dynamically. Two particular devices to be discussed in more detail are the seismic instrument and the piezoelectric crystal. Remember that a very wide variety of sensors is possible. Precision and dynamic-response capabilities have made these latter two devices very prominent in the field.

The basic elements of a seismic instrument are shown in Figure 7.38.

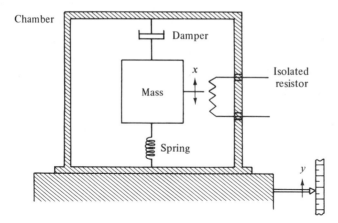

Figure 7.38 Seismic Instrument

The mass is protected by a chamber and supported by a spring. A damper is provided to control the dynamic operation to a desired range. The position of the mass relative to the chamber is measured by any one of many possible systems already discussed. The basic equations governing this system have already been presented.

$$M\frac{d^2x}{dt^2} + C\frac{d(x-y)}{dt} + k(x-y) = 0 \tag{7.14}$$

where $M =$ seismic mass, $C =$ damping constant [damping force $=$ C(relative velocity)], and $k =$ spring constant. Since the system senses relative motion between the seismic mass and the chamber, Md^2y/dx is subtracted from both sides of the above equation to obtain

$$M\frac{d^2(x-y)}{dt^2} + C\frac{d(x-y)}{dt} + k(x-y) = -M\frac{d^2y}{dt^2} \tag{7.15}$$

Therefore, this system responds to an input acceleration. Let $x - y = z$; then dividing both sides of Equation 7-15 by M gives

$$\frac{d^2z}{dt^2} + \frac{C}{M}\frac{dz}{dt} + \frac{kz}{M} = -\frac{d^2y}{dt^2} \tag{7.16}$$

The dynamic response of second-order systems to various forcing functions is presented in Appendix III. However, this particular application requires some special attention. Sensors have already been discussed which could measure either z or dz/dt. Therefore, the output signal from this measurement system would represent one of these quantities. How can these measured quantities be related to the input acceleration? This question can only be answered by looking at several types of input accelerations. First, let us consider a constant acceleration.

If d^2y/dt^2 is equal to a constant, the solution of Equation 7.16 for steady state gives

$$z = -\frac{M}{k}(\text{const}) \tag{7.17}$$

The total solution will also include the transient or homogeneous solution of

$$\frac{d^2z}{dt^2} + \frac{C}{m}\frac{dz}{dt} + \frac{k}{m}z = 0$$

or

$$z = e^{-(C/2M)t}(A\cos\omega t + B\sin\omega t) \tag{7.18}$$

where

$$\omega = \sqrt{\frac{k}{m} - \left(\frac{C}{2m}\right)^2}$$

and A and B are constants determined from initial relative position and velocity.

When there is no damping $(C = 0)$, the spring mass has a frequency of oscillation equal to $\sqrt{k/m}$, which is called the natural frequency of the system. When $C = 2\sqrt{k/m}$, the sine–cosine terms vanish in the homogeneous solution; this is called critical damping. Figures involving these parameters are given in Appendix III. In any case of finite damping, the homogeneous solution is decreased to zero because of the negative exponential coefficient. Therefore, the total solution for constant acceleration,

$$z = e^{-(C/2m)t}(A\cos\omega t + B\sin\omega t) - \frac{M}{k}(\text{const}) \tag{7.19}$$

approaches the constant value, $-M/k(\text{const})$, after some transient time interval. For constant acceleration the measured output eventually becomes a constant times the input acceleration.

If the input acceleration varies linearly with time $(d^2y/dt^2 = At + B)$, the transient solution is unaffected by any input signal and the steady-state solution becomes

$$z = -\frac{AM}{k}t + \frac{MB}{k} - \frac{CAM}{k^2} = -\frac{M}{k}(At + B) + \frac{CAM}{k^2} \tag{7.20}$$

After some transient time, the output signal, z, is equal to a constant times the input signal plus another constant:

$$I_o = (\text{const})_1\,(I_i) + (\text{const})_2$$

This situation is acceptable for a measurement system, since the constants can be easily established.

Finally, if the input signal is a sine wave, both the natural frequency and the damping coefficient are involved in the final output signal. If the damping coefficient is selected to be 0.65–0.70 times the critical damping coefficient, the output signal is approximately equal to a constant times the input acceleration for input frequencies less than one-half the natural frequency of the seismic instrument (see Appendix III). Using a transducer to measure the relative velocity for the output signal with the same instrument would give output relative velocity equals a constant times the input velocity when the input frequency is approximately three times the natural frequency of the system. The output relative displacement represents the input relative displacement for the same frequency ratio. Therefore, the basic seismic device is capable of measuring displacement, velocity, and acceleration of a sinusoidal input signal. The relative-displacement transducer measures acceleration for input frequencies less than one-half the natural frequency and measures displacement for input frequencies greater than three times the natural frequency. Velocity sensors could measure input velocity for input frequencies three times the natural frequency.

The seismic mass is usually designed for relatively low-frequency input signals (less than 300 Hz). To extend the range for higher-frequency input signals, another type of accelerometer is required. The piezoelectric crystal is one sensor that satisfies the high-frequency requirement. It senses the force required to accelerate a body, and an electrical charge is induced on the crystal. Input frequencies below 10Hz experience some distortion, but at higher frequencies the output is linear with input acceleration up to 10,000 Hz. At high acceleration values [up to (10,000)(earth's gravity)], the crystal operates very nicely. At lower acceleration (below 10 g) the temperature of the crystal must be carefully controlled. Acceleration and frequency ranges do affect the accuracy of the crystal. However, it is very versatile and relatively inexpensive.

ANGULAR ACCELERATION

The measurement of angular acceleration actually presents problems only in mounting and in accounting for centrifugal accelerations. The previous systems for measuring acceleration (or force) can be adopted to measure angular acceleration.

The piezoelectric crystal may be sensitive to both the angular acceleration and the centrifugal acceleration when placed at the edge of a rotating disc. If the angular velocity is measured simultaneously, a correction for the centrifugal force can be made.

If the housing for a seismic mass is designed to rotate with a shaft, a torsional spring can attach the mass to the housing. Then the moments of inertia of the mass, the damping fluid, and spring constant replace corresponding terms in the previous seismic mass analysis:

$$I\frac{d^2(\theta - \phi)}{dt^2} + C\frac{d(\theta - \phi)}{dt} + K(\theta - \phi) = -I\frac{d^2\phi}{dt^2} \qquad (7.22)$$

where I = moment of inertia of the seismic mass
  C = damping coefficient
  K = spring constant
  θ = rotation of seismic mass
  ϕ = rotation of housing

All seismic masses are sensitive to accelerations in any direction in addition to the one direction for which it was designed.

7.6 Other Parameters and Time Systems

The measurement of any combination of parameters with time involves an evaluation of the dynamic response of the sensor system except when steady-state operation is being considered. Since energy conversion and transmission is a very common engineering problem, the measurement of time-dependent parameters is essential. Mass and volume flow rates and transient temperatures and pressures will be discussed here. The approach is to present the sensors' operating principles, discuss calibration, and indicate limitations. More comprehensive details can be found from a list of NBS publications.

MASS AND VOLUME FLOW RATES

The problems inherent in measuring velocity are also present in the measurement of mass and volume flow rates. If the density of the flowing material is known, mass and volume are related. Therefore, the measurement of one determines the other. We will discuss mass flow rates with the understanding that density is known.

Probably the simplest method for measuring mass flow rate is to collect the mass flowing for a measured time interval and use a mass balance. If the mass is liquid and the flow is steady, a bucket and timer can provide the most reliable and accurate measure of mass flow rate available. This method is used to calibrate other flow-rate systems. This method is directly traceable to the basic standards of Chapter 6. If the mass flowing is air, an empty plastic sack of known (usually large size) volume can be used to find volume flow rate. This may at first seem impractical, but in many cases it is the most effective method.

A second class of meters for measuring mass flow rate is the obstruction meters. These can be calibrated by the above systems and used for either air or water. Calibration for water flow rates can be used for air in most cases by application of the principles of similarity of flow. When gas flow rates approach compressible flow conditions (velocity of flow greater than 0.3 sonic velocity) the effect of compressible flow must be considered. Reference 11 gives some measurement background in the wide variety of flow-meter requirements that are being met. A very general source of information on measurement methods is given in reference 12. This reference particularly introduces methods for some obstruction meters and for pitot-tube measurement.

Six different flow meters are shown in Figure 7.39. The first three are basically dependent on the Bernoulli equation for their operation.

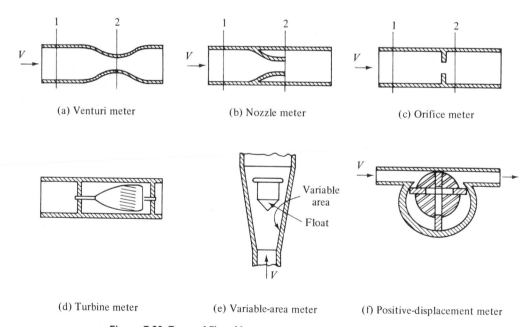

(a) Venturi meter   (b) Nozzle meter   (c) Orifice meter

(d) Turbine meter   (e) Variable-area meter   (f) Positive-displacement meter

Figure 7.39 Types of Flow Meters

$$P_1 + \tfrac{1}{2}\rho_1 V_1^2 + \rho_1 Z_1 = P_2 + \tfrac{1}{2}\rho_2 V_2^2 + \rho_2 Z_2 \qquad (7.23)$$

The venturi, orifice, and nozzle meters are obstruction meters. They are operated in the horizonal plane to eliminate the effects of elevation, Z, and the temperature is assumed to be constant. If the static pressures are measured at points 1 and 2, an incompressible fluid ($A_1 V_1 = A_2 V_2$) will have the velocity determined by

$$V_2 = \frac{1}{\sqrt{1 - (A_2/A_1)2}} \sqrt{\frac{2}{\rho}(P_1 - P_2)} \qquad (7.24)$$

The ideal volume flow rate would be $Q_{ideal} = A_2 V_2$. The actual flow rate varies with Reynold's number and with the individual meter. Then,

$$\frac{Q_{actual}}{Q_{ideal}} = C \qquad (7.25)$$

where the discharge coefficient, C, is dependent on the above parameters. Values of C for various meters is given in references 12 and 13. Local calibration should also be performed using collection tanks and timers to be sure that the design and instrumentation of the meter meets the specifications. When the flow is compressible, an additional correction factor, the expansion factor, must be included in the analysis of the meter.

The turbine meter of Figure 7.39(d) has gained recognition as a precision flow-measurement device (references 1 and 11). The rotational speed of the turbine is dependent on the velocity at the turbine blades. The output may be a magnetic pick-up to isolate the turbine from the housing. Accuracies better than ± 0.2 per cent of the reading have been obtained for particular turbine meters.

The variable-area meter of part (e) has a restriction in the form of a float which is slightly more dense than the fluid flowing. The pressure difference across the float is a result of the fluid drag on the float. This drag force is balanced by the gravitational attraction of the float. When the velocity increases, the height of the float from the inlet increases. A scale is etched on the glass variable-area chamber. This meter requires a housing that is transparent, or some position sensor must be attached internally in the system. The range of this meter is limited by the geometry and densities of fluid and float. Some meters come with several different density floats.

The positive-displacement meter shown in part (f) will sweep out a constant volume during each revolution. A measure of the rotational speed determines the volume flow rate if the chamber is always full of the fluid flowing. The internal sliding vanes are spring loaded so that they always extend to the housing. Some leakage may still be possible for this system. However, the system is relatively reliable and certainly merits consideration for many applications.

All of the systems discussed to this point have been average-value flow-rate sensors. Although it is not possible to measure exactly the momentum of a particular molecule, several much more local fluid-velocity measurement systems are possible. Knowing the velocity profile in a conduit and the density allows for the determination of mass flow rate. One such device is the pitot tube shown in Figure 7.40. When the pitot tube is used to measure

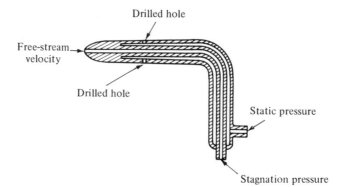

Drilled hole

Free-stream velocity

Drilled hole

Static pressure

Stagnation pressure   **Figure 7.40** Pitot Tube

the velocity in an incompressible fluid, the difference between stagnation pressure, P_t, and static pressure, P_s, represents the free-stream velocity

$$P_t - P_s = \tfrac{1}{2}\rho V_\infty^2 \tag{7.26}$$

When the fluid flowing is compressible, the ratio of total isentropic stagnation pressure to static pressure of an ideal gas is given by

$$\frac{P_t}{P_s} = \left(1 + \frac{\gamma - 1}{2} M_\infty^2\right)\frac{\gamma}{\gamma - 1} \tag{7.27}$$

where $M_\infty = V_\infty/a$ is the free-stream Mach number, $a = \sqrt{\gamma RT}$ is the acoustic velocity in the fluid, and γ is the isentropic exponent of an ideal gas ($P_1 V_1^\gamma = P_2 V_2^\gamma$).

Since the pitot tube is a very common sensor, some comments will be made concerning this sensor. The tube must be perfectly aligned with the axis of flow to provide the best total pressure and to avoid sensing a velocity component in the static-pressure drill holes. The static-pressure taps must be located approximately five outside tube diameters downstream from the probe point to allow the static pressure to regain its initial value. The probe disturbs the flow field (first axiom of measurement), and wakes influence the static pressure down stream. Compressible fluids may not be ideal gases. Steady flow is required for proper operation. Reference 12 gives a testing code for determining the velocity profile and the corresponding mass flow rate. The use of a pitot tube in supersonic flow ($M_\infty > 1$) requires more analysis (reference 14). At very low flow rates the pitot tube may be somewhat erratic, with decreased accuracy,

The hot-wire anemometer uses the fact that the heat-transfer rate away from a cylindrical wire depends on the velocity of a gas across the wire under forced convection. The thermal conductivity of gases is small, so the heat-transfer rate is greatly enhanced by gas velocity. Liquids could also be measured, with somewhat less sensitivity. A hot-wire anemometer probe is shown in Figure 7.41. The sensor-wire diameter may be decreased to

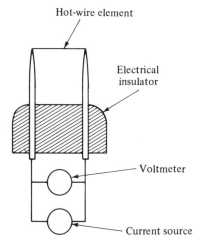

Figure 7.41 Hot-Wire Anemometer Probe

increase the sensitivity of the device. With extremely small wire filaments, the turbulent components of fluid flow may be sensed. Two types of instrumentation are used with this probe. A constant current can be applied to the wire; then the wire resistance is a function of the wire temperature, and a measurement of the voltage drop across the wire can be used with the constant current to determine this resistance. This method is generally less expensive than the alternative method of maintaining a constant temperature of the wire (also constant resistance) by varying the current flow. The current is measured to determine flow. The constant-temperature control system is subject to feedback problems, especially when electrical noise is present. It requires more experience by the operator, and the resolution may not be as good as that of the constant-current device at high frequencies.

In both cases the filament is very fragile, and the system should be calibrated at frequent intervals because the wire properties can easily change with use and exposure to high temperatures. The filament may also be easily burned out if a constant current is applied and the flow rate decreases. Electrical noise is a serious extraneous input, and the form of this signal may be similar to the turbulent fluid flow sensed by the filament. This type of sensor has enjoyed much success because it provides turbulent flow data. The same principle of operation could be applied using an extremely small thermistor to measure temperature and to provide heat, or by using a thermocouple junction attached to the hot-wire filament. The operation would be the same, but the method of determining temperature would be altered.

Several optical methods for measuring local velocity have been very successful. These systems will be discussed in Chapter 9. They do offer some basic advantages not available to other sensors.

TRANSIENT TEMPERATURE AND PRESSURE

Problems in the mass, momentum, and energy-transport area could require experimental information on the variation of temperature and pressure with time. The sensors for measuring these parameters have already been introduced. Again Appendix III is suggested for more details about dynamic calibration. However, the measurement systems will be discussed here relative to their dynamic capabilities.

If a transient temperature occurs in less than $\frac{1}{10}$ s, many of the sensors discussed earlier would not respond to the signal. Almost all sensors must be heated to the temperature of the medium before they can produce an output signal of the proper magnitude (the optical pyrometer is one exception). The heat-transfer rate to and from the sensor governs the basic response,

$$Q = MC\frac{dT}{dt} \tag{7.28}$$

where Q = heat-transfer rate
   M = sensor mass
   C = specific heat of mass

T = temperature of mass

t = time

If the heat-transfer rate is independent of time and temperature, the temperature of the mass varies linearly with time. However, the heat-transfer rate is almost always dependent on sensor temperature. All modes of heat transfer to and away from the sensor must be included in an analysis. The sensor generally has a first-order response to a change in temperature,

$$(\tau D + 1)I_o = I_i \tag{7.29}$$

For these cases the smaller mass gives the faster response. The sensors are limited by the "thermal inertia" of their mass. The anemometer (resistance wire) is one sensor for measuring dynamic temperature. The thermocouple and thermistor may both be miniaturized to give high response. The dynamic response is generally limited by the sensor size, because the output signal is in the form of an electrical property. Therefore, the intermediate modifying elements and the read-out device can respond to signals above 20,000 Hz. The development of thin-film thermocouples and thermistors has reduced the thermal inertia of the sensor to a value in the range of electrical responses. However, the most commonly used sensors may be assumed to be a first-order system connected to two zeroth-order systems to give an output of the total measurement system that is a first-order device.

Let us place a thermocouple in a stream of gas at a temperature T_g. We will also assume that the forced-convection heat-transfer coefficient is a constant, h. Then

$$Q = hA_T(T_g - T_T) = M_T C \frac{dT_T}{dt} \tag{7.30}$$

where the subscript T is the thermocouple junction.

If the junction is assumed to be spherical,

$$\frac{M_T}{A_T} = \frac{\frac{4}{3}\pi r^3}{4\pi r^2} = \frac{r}{3}$$

Equation 7.30 becomes

$$\frac{3h}{rc} = \frac{1}{(T_g - T_T)} \frac{dT_T}{dt} \tag{7.31}$$

Solving this gives

$$T_T = (T_g - T_{T0})(1 - e^{-3ht/rc}) + T_{T0} \tag{7.32}$$

where T_{T0} is the initial temperature of the thermocouple. The dimensionless quantity plotted in Appendix III for a step change is

$$\frac{T_T - T_{T0}}{T_g - T_{T0}} = (1 - e^{-t/\tau}) \tag{7.33}$$

where $\tau = hA_T/M_T C = 3h/rC =$ the parameter in Equation 7.29.

Two basic types of dynamic pressure sensors are the piezoelectric crystal and the flexible diaphragm. Both can, theoretically, respond to frequencies above 20,000 Hz. Calibration of these sensors could be accomplished in a shock tube. Several methods are possible for use with the diaphragm. The strain gage on a diaphragm has a response essentially limited by the time required for the diaphragm to deform under the changing pressure. Both the piezoelectric crystal and the strain gage on diaphragm are zeroth-order instruments if the input pressure frequency is less than 1000 Hz. They may be capable of responding to higher-pressure frequencies, but they should be calibrated before use in this range. The 1000-Hz limit is suggested only to indicate that more complete and careful calibration should be made when higher frequencies are expected.

Dynamic temperature and pressure sensors have been developed rapidly in the last few years. Optical sensors have been used in this field, and they will probably provide some additional capabilities for dynamic measurements. It should be pointed out that optical systems can be zeroth-order systems for dynamic response. They are limited only by the speed of light. However, the intermediate circuitry and read-out devices may actually be the limiting factor for optical sensors. It should also be mentioned that a dynamic pressure sensor could be modified to measure dynamic force.

7.7
Simple
Electrical
Parameters

It has been necessary to introduce the measurement systems available for measuring resistance at earlier points in this book. This material could also be included here, but we will only refer the reader to the discussion of the Wheatstone bridge, the Mueller bridge, and the Kelvin bridge of Chapter 6. Two additional electrical parameters that are commonly found in measurement systems are capacitance and inductance.

According to reference 15, the standard ohm may be determined from a calculable capacitor. Only the accuracy of the permeability of space associated with the speed of light accuracy and the accuracy of measuring the local acceleration of gravity limit the accuracy of calculating the value of a standard capacitor. This then would establish a value for the absolute ohm. Presently, the absolute ohm is established from a defined inductance through a very time-consuming experimental program. The fact that the ohm is defined by the parameters capacitance and inductance means that they can be calculated mathematically and determined experimentally with great accuracy (approximately one part in 10^7).

Standards for measuring capacitance and inductance are available for use in local laboratories. These may be calibrated by the NBS (reference 15). Both parameters vary with the frequency of the input signal, and a variety of bridge circuits are available for local calibration and measurement. An adequate bibliography on the measurement of all electrical parameters is given in reference 15. If the coverage here is not complete enough for a particular application, it is recommended that this reference be used. Several bridge circuits are available for measuring either capacitance or inductance. Only a few of them are discussed here. Figure 7.42 shows four bridge circuits that are used in measuring capacitance and inductance.

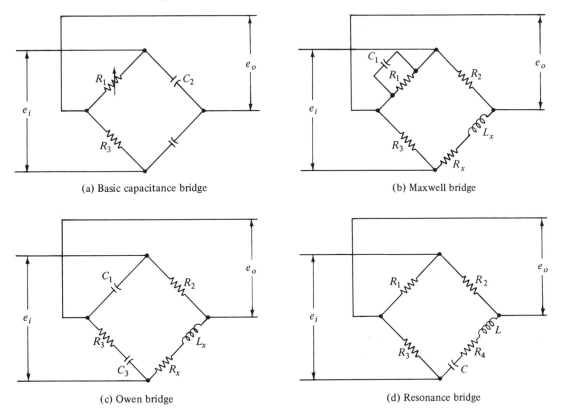

(a) Basic capacitance bridge

(b) Maxwell bridge

(c) Owen bridge

(d) Resonance bridge

Figure 7.42 Four Inductance and Capacitance Measuring Bridge Circuits

Part (a) is a basic capacitance bridge. When the output voltage is adjusted to zero by changing resistor R_1, the unknown capacitor is determined in terms of one known capacitor, C_2, and two known resistances. (R_1 must be measured after null output has been obtained.)

$$C_x = C_1 \frac{R_1}{R_2}$$

If a known inductor and an unknown inductor are used to replace the capacitors,

$$L_x = L_2 \frac{R_1}{R_3}$$

This bridge circuit is very simple to operate and requires only one standard capacitor or inductor for one given range of unknown parameters.

In part (b) an unknown inductance can be measured in terms of a known capacitor and three known resistors. (Again the resistors may be variable but must be measured after null balance is achieved.) Both of these bridges are independent of the frequency of the exciting voltage. They both have only one unknown reactive element, and resistors are the only variables parameters required for balance. And changes in resistor R_1 affect only the resistive balance of the Maxwell circuit. The conditions for null balance are

$$L_x = R_2 R_3 C_1$$

$$R_x = \frac{R_2 R_3}{R_1}$$

The Owen bridge of part (c) requires two known capacitors plus known resistors to measure an unknown inductance. For null balance

$$L_x = R_2 R_3 C_1$$

$$R_x = \frac{R_2 C_1}{C_3}$$

This bridge is also independent of frequency.

The bridge of part (d) depends on the exciting voltage frequency. If the input frequency is known and one reactive element is known, the bridge can be used to determine the other reactive element. Null balance is achieved when resonance occurs:

$$X_L = X_C$$

This is obtained by adjusting the voltage frequency and one resistive element. When resonance occurs,

$$R_4 = \frac{R_2 R_3}{R_1}$$

$$\text{frequency} = \tfrac{1}{2}\pi\sqrt{LC}$$

Other resonance circuits detect either a maximum or a minimum output voltage. These generally have less accuracy than the resonance bridge circuit discussed here.

The measurement of dielectric properties and magnetic properties will not be discussed here. Reference 15 is a good source for investigating measurement systems for these electrical properties. Electrical transformers are also discussed in reference 15. Anyone primarily concerned with the measurement of electrical properties should obtain a copy of this reference. It is low in cost and is a valuable source for electrical measurement systems.

7.8
Conclusions   The basic dimensions defined in Chapter 6 were the foundation for all of the parameters introduced in this chapter. Any extraneous input which affects the basic dimension also affects the derived dimension. The basic accuracy of the derived dimension is established by the accuracy of the primary standard for each basic dimension incorporated in the new dimension.

The possibility of defining a new parameter independently from the basic dimensions was explored for the parameter, force. For a new parameter to be consistent with the basic dimensions and with the physical laws, a dimensional constant must be determined for each physical law. The complexity introduced by this action is demonstrated by the widespread lack of understanding of the dimensional constant $g_c = 32.174 \text{ ft-lb}_m/\text{lb}_f\text{-s}^2$. Fortunately, most other new dimensional parameters have been defined in terms of the basic dimensions. The constraints imposed on all dimensional parameters by the basic dimensions should be clearly understood.

The problems of calibrating systems for the measurement of derived parameters are generally solved by using the technology developed by the NBS. They provide secondary-standard calibration and certification services. It is usually much easier to depend on their efforts than it is to attempt to develop a local technology for providing this calibration.

Dynamic calibration is a problem considered in Appendix III. It involves establishing a dynamic input signal that can be traced to a basic standard.

Mechanical systems are limited in dynamic response by inertia and friction. Electrical systems are generally capable of responding to much higher frequency input signals than are mechanical systems. Dynamic calibration should involve standard signals that are similar to the input signal they are to measure.

The calibration problem generally increases with the complexity of the derived parameter. The derivations in this chapter involved the least complex combinations of basic dimensions. The parameters considered in the next chapter will be more involved and will require more careful attention to the extraneous inputs. As discussed in the previous chapter, change in temperature causes a variety of errors in the measurement of mass. The mass balance arms change in length, the mass volume changes, and the air density changes. In the present chapter the measurement of density was seen to involve all of these errors. It also involves changes in volume of the container in which a liquid to be measured is placed. This introduces one more possible extraneous input. One must be alert to all possible extraneous inputs.

At this point the logical development of measurement systems should be cultivated. The basic factors that influence sensor selection, range, accuracy, dynamic response, and extraneous inputs should be clearly understood. The goals of an experimental program are influenced by the measurement system, calibration, and data analysis. For a program to be successful the role of each part should be reviewed.

References

1. *Precision Measurement and Calibration—Heat and Mechanics*, NBS Handbook 77, vol. 2, U.S. Department of Commerce, Washington, D.C., 1961.

2. BECKWITH, T.G., and BUCK, N.L., *Mechanical Measurements*, 2nd ed., Addison-Wesley Publishing Co., Reading, Mass., 1969.

3. PERRY, C.C., and LISSNER, H.R., *The Strain Gage Primer*, 2nd ed., McGraw-Hill Book Co., Inc., New York, 1962.

4. "Instruments and Control Systems," *Pressure Handbook*, Rimbach Publications, Pittsburgh, Pa., n.d.

5. *Handbook of Measurement and Control Instruments*, Rimbach Publications, Pittsburgh, Pa., 1959.

6. "Instruments and Control Systems," *Vacuum Measurement Handbook*, Rimbach Publications, Pittsburgh, Pa., 1966.

7. MANGOLD, V.L., *Pressure Measurement in the* 10^{-4} *to* 10^{-10} *Torr Range*, Technical Report AFFDL-TR-69-49, (AD-700983).

8. BROMBACHER, W.G., *A Survey of Ionization Vacuum Gages and Their Performance Characteristics*, NBS Technical Note 298, U.S. Department of Commerce, Washington, D.C., 1967.

9. TATE, P.R., *Gravity Measurements and the Standard Laboratory*, NBS Technical Note 491, U.S. Department of Commerce, Washington, D.C., 1969.

10. HILTON, J.S., *Accelerometer Calibration with Earth's Field Dynamic Calibrator*, NBS Technical Note 517, U.S. Department of Commerce, Washington, D.C., 1970.

11. COTTON, K.C. (ed.), *Flow Measurement Symposium*, American Society of Mechanical Engineers, New York, 1966.

12. *ASHRAE Guide and Data Book, Fundamentals and Equipment for 1965 and 1966*, ASHRAE, Inc., New York, 1965.

13. *Fluid Meters, Their Theory and Applications*, 5th ed., American Society of Mechanical Engineers, New York, 1959.

14. HOLMAN, J.P., *Experimental Methods for Engineers*, McGraw-Hill Book Co., Inc., New York, 1966.

15. *Precision Measurement and Calibration, Electricty—Low Frequency*, NBS Special Publication 300, vol. 3, U.S. Department of Commerce, Washington, D.C., 1968.

PROBLEMS

7-1. Let area be given a defined dimension, 1 circle (where 1 circle is the area of a 1-ft-diameter circle). What dimensional constant is required by the physical law: 1 circle = K(length in feet)^2?

7-2. If the density of 1 gal of water at 70°F is defined to be 1 blob, what dimensional constant must be found to relate 1 blob to the basic dimensions for mass and length?

7-3. If the force of attraction between the earth and a 1-lb mass at a location where the acceleration of gravity is 32 ft/s^2 is defined to be 1 lb$_f$, what dimensional constant is necessary to relate this new dimension to the basic dimensions?

7-4. Design a system for measuring force. Tell how it could be calibrated. Predict its dynamic response, static sensitivity, range, and accuracy.

7-5. List 10 possible extraneous inputs to a bathroom scale.

7-6. Is it possible to measure the strain in a metal shaft at a point? If the answer is yes, explain the loading conditions under which this could be accomplished.

7-7. List 10 possible extraneous inputs to a mounted strain gage.

7-8. Tell how the accuracy of a planimeter could be checked. Give complete details.

7-9. Tell how the pressure at a point could be measured. If this is possible, give the conditions that must be maintained.

7-10. Based on the accuracy possible from basic dimensions, what is the maximum possible accuracy with which pressure can be measured?

7-11. How does the maximum possible accuracy of a stress measurement compare to the pressure measurement of Problem 7-10?

7-12. If the weights for a dead-weight tester are accurate ± 0.01 lb_f, and the piston diameter is accurate ± 0.001 in.^{2}, what is the accuracy of the pressure? (Assume one standard deviation for all accuracy data.)

7-13. Repeat Problem 7-10 for (a) density, (b) volume, (c) torque, (d) acceleration.

7-14. Give a calibration technique for rotational speed. Apply this to the development of a linear-velocity calibrator.

7-15. Give one set of conditions where the instantaneous velocity could be measured. What measurement system would be required?

7-16. List 10 systems for measuring acceleration.

7-17. What extraneous inputs are possible in the scale and stop watch method of measuring mass flow rate? Pitot-tube Anemometer?

7-18. List the factors that affect the accuracy of each of the measurement systems of Problem 7-17.

7-19. Is a thermocouple always a first-order instrument? What factors determine the order of a measurement system? (See Appendix III.)

7-20. Is the seismic instrument always a second-order system?

7-21. Explain how a soft drink bottle could be calibrated to measure the density of a liquid if a mass balance is available.

7-22. How does the error in measuring resistance and the error in a standard capacitor affect the error in measuring an unknown capacitance? (Give an answer based on one of the bridge circuits.)

7-23. How can the variable resistor in the bridge circuits of Section 7.7 be measured accurately? Give all details.

7-24. Is dynamic calibration necessary for a system that is to be used to measure a static signal?

7-25. Is the piezoelectric crystal a zeroth-order sensor? If it is, are there input frequency ranges over which it is not a zeroth-order sensor?

7-26. Can a piezoelectric crystal measure acceleration? If it can, explain why it is capable of sensing the signal.

7-27. How could a barbed-wire fence be used to measure weight?

7-28. How does the measurement of the speed of light and the local acceleration of gravity influence electrical properties?

8

A SURVEY OF OTHER DIMENSIONAL PARAMETERS

8.1
Introduction

Every physical property is related to the basic dimensional system that is initially defined. The methods for developing measurement systems for any parameter are all based on the governing physical laws. It would be impossible to cover every possible physical property in one book. Therefore, only a selected group of additional parameters will be considered. Which ones to present is always a difficult decision to make; one would like to present measurement techniques from all fields. The background of the author always biases the choices of parameters that will be discussed. It is hoped that the choices made here will be adequate for those interested in the thermal sciences.

If a particular area of interest requires different types of measurement systems, they can be considered in the same format that is presented here. First, the physical problem is to be defined and the governing equations are to be determined. Then, we will look for sensors capable of measuring the associated variables necessary to evaluate the parameter under consideration. Many of the components required have already been discussed; we have many building blocks at our disposal. Finally, the analysis of the overall measurement system for accuracy will be attempted. This is the same step-by-step procedure that should be followed in the development of any area of measurement.

8.2
Work and
Power

All energy-transport problems involve the measurement of the work or the power (work per unit time) that is transported into the system under analysis. Work is defined in thermodynamics to be that energy-transport process across the boundary of a system which is equivalent to a force acting on a mass and moving it a distance. This basic concept of work is expressed mathematically in the form of line integral,

$$W = \int_c \vec{F} \cdot \vec{dS} \tag{8.1}$$

where $W = \text{work}$
        $\vec{F} = \text{the vector force}$
        $\vec{dS} = \text{the vector position along curve } c$

If the displacement occurs in the vertical direction only, and the force is the local gravitational attraction, Equation 8.1 allows work to be measured by measuring the weight of a mass (mg/g_c) and the change in vertical position. The basic equivalence of work is the lifting of a weight. Force and length sensors have already been discussed, and the error associated with measuring mass is

$$dW = F\,d(\Delta x) + \Delta x\,dF \tag{8.2}$$

If the errors in measuring Δx and F are normally distributed and expressed in terms of one standard deviation, the root-sum-square standard deviation in measuring work is

$$\Delta W_{rss} = \sqrt{[F\Delta(\Delta x)]^2 + [(\Delta x)(\Delta F)]^2} \tag{8.3}$$

If the time interval (Δt) over which the work is done is known, the power is $P = W/\Delta t$. For a normally distributed error in measuring Δt and one standard deviation limits,

$$\Delta P_{rss} = \sqrt{\left[\frac{\Delta W_{rss}}{(\Delta t)}\right]^2 + \left[\frac{-W}{(\Delta t)^2}\right]^2} \tag{8.4}$$

The error in measuring power involves one term more than that for work.

Electrical work is generally assumed to be a pure form of mechanical work. The assumption that they are equivalent is valid if one form of work can be entirely converted to the other form. Measurements prove that they are not absolutely equivalent, but the assumption can be made with small error. Electrical power is a function of the voltage applied and the current flowing,

$$P = EI \tag{8.5}$$

where P = electrical power
          E = voltage
          I = current

For alternating current, both E and I are functions of time, and the work done during a time interval $t_2 - t_1$ is

$$W = \int_{t_1}^{t_2} E(t)I(t)\, dt \tag{8.6}$$

Usually, voltage and current are sine waves with the same frequency and with a phase difference. For this case the power can be found by finding the work for one cycle,

$$P = \frac{1}{T} \int_0^T (E \sin \omega t)(I \sin \omega t + \phi)\, dt = \frac{EI \cos \phi}{2} \tag{8.7}$$

where $T = 2\pi/\omega$
          ω = frequency of the signal
          E = voltage amplitude
          I = current amplitude
          ϕ = phase angle between the signals

We already have methods for measuring voltage and current. The phase relationship can be obtained from Lissajous patterns. If they are in phase, Equation 8.7 is the product of the root-mean-square voltage times the root-mean-square current,

$$E_{\text{rms}} I_{\text{rms}} = (0.707E)(0.707I) = 0.5EI$$

Since electrical properties were defined independently of the basic dimensions, a dimensional constant is required to obtain dimensional consistency, 1 watt(W) $\doteq$ 2654.16 ft-lb$_f$/h. The errors in measuring electrical power and work may be obtained using the same technique as that for mechanical properties. However, power is measured directly, and work is obtained by multiplying by a measured time.

The testing of power equipment has long been one goal of a measurement system. The American Socity for Testing Materials (ASTM) has developed various test codes for different types of power units. One test code for evaluating fan performance was discussed in Chapter 4. Work could be determined from the lifting of a weight, the generation of electrical power, the movement of a mass of fluid, or the dissipation of energy into internal energy changes. The analysis of rotating-shaft power output is usually obtained from torque dynamometers. These have already been discussed in Chapter 7. If the torque and rotational speed are measured, the power can be determined from the relationship

$$P = 2\pi T\omega \tag{8.8}$$

where $T =$ torque

$\omega =$ rotational speed in revolutions per unit time

In all cases, part of the energy of the shaft is dissipated by friction. This dissipated energy is classified as heat loss. One must perform a complete energy balance to obtain a measure of the potential work output of a shaft. One such analysis is presented in Chapter 4. It should be obvious that extraneous inputs and more complex physical relationships require careful analysis of the design of a measurement system.

8.3
Heat

Heat is another thermodynamic concept. It is all energy that is transported across a system boundary except that of mechanical or electrical work. Heat energy in transit is essentially measured by determining the net effect it has on a thermodynamic system. This means that one must be able to measure enough thermodynamic properties to determine the thermodynamic state of the system. Density, pressure, and temperature are three thermodynamic properties that have already been discussed in terms of measurement. The thermodynamic system of Figure 8.1 allows no mass

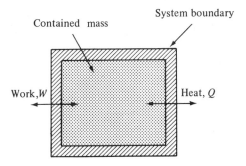

Figure 8.1 Thermodynamic System

to be transferred across the boundary, but it does allow heat and work to be transferred either into or out of the system. The first law of thermodynamics is the conservation of energy. It states that all of the energy transported into the system is either stored or transported out of the system:

$$W_{in} + Q_{in} = \text{energy stored} + W_{out} + Q_{out} \tag{8.9}$$

The energy stored is in the form of the internal-energy change of the mass inside the system boundary and the internal-energy change of the boundary material. The internal energy of several fluids and several solids has been obtained by previous investigators. It is not always clear what the accuracy of this data is. However, it is generally assumed that these internal-energy properties are accurate to the number of significant figures plus or minus one-half of the last significant figure. This assumption will usually be made until some experimental evidence to the contrary is obtained. Thermodynamic properties of air and water are given in several sources. They are commonly used in energy-analysis experiments. Specific heats for various alloys of aluminum and steel are also available in graphical and tabular form. Experimental methods for determining these and other thermal properties will be discussed in this chapter.

If we could insulate the system of Figure 8.1 so that no heat was allowed to cross the boundary, work in the form of electrical energy could be transported into the system and dissipated in an electrical resistor. Measurement of initial mass, temperature, and pressure could be used to determine the change in internal energy of the system. This would constitute an initial calibration of the system. (This would also require zero work flow out of the system.) The measurement of thermodynamic properties, pressure and temperature, would establish the thermodynamic state for a single-component single-phase fluid. The energy stored in the boundary material could then be calibrated. One extraneous input to this analysis is the assumption that zero heat is transported across the boundary. This assumption can be checked if the thermal capacity of the container and its temperature change is known. Otherwise, the heat loss will be collected into the energy stored in the boundary material.

After the above calibration has been completed, the work input can be set equal to zero and the system can be connected to a heat source (or sink) to measure the transport of heat across the boundary. Usually, heat transport by conduction is used. Mathematically,

$$Q = -KA\frac{\Delta T}{\Delta x} \tag{8.10}$$

where $K = $ thermal conductivity of a solid

A = cross-sectional area of the solid perpendicular to the flow of heat

$\dfrac{\Delta T}{\Delta x}$ = the temperature gradient in the direction of heat flow

If we measure temperature, position, and area, the system of Figure 8.2

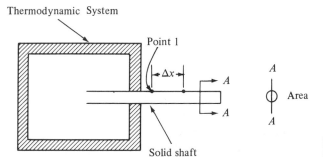

Figure **8.2** Thermal Conductivity Measurement

can be used to measure thermal conductivity. If any heat is transferred away from the solid shaft between point 1 and the system boundary, the thermal conductivity will be in error. One way to reduce this error is to place the entire system in an evacuated space to reduce convection heat losses and to place radiation shields around the shaft to reduce radiation heat losses. A less desirable solution is to cover the shaft with a very-low thermal conductivity material such as fiber glass. Extraneous inputs are almost impossible to eliminate completely. We simply attempt to predict the errors associated with various physical situations.

Now we have a theoretical method for measuring heat flow, thermal conductivity, heat capacity (also called specific heat), length, area, or temperature difference. Length, area, and temperature are much more accurately measured by other methods, but this system is sensitive to these parameters. The accuracy associated with the system of Figure 8.2 is limited by the heat-loss problem. Reference 1 devotes several articles to methods for reducing heat loss. The common name for the system discussed here is the calorimeter. More details of the use of calorimeters and of the subsequent data analysis are obtainable from reference 1.

One other application of the calorimeter is in determining the heat of combustion of different fuels. The energy generated from a chemical reaction is a result of changing the chemical energy of a mass into the internal energy of its products. Chemical energy may be thought of as a form of the internal energy of a mass. The energy given up in combustion will be dissipated in a manner identical to that of heat flow into the system. An equivalence between chemical energy and heat transfer is, therefore, possible using the calorimeter discussed. We could calibrate the system using electrical heaters and then place the chemically reacting material in the system. After combustion, the measurement of the thermodynamic properties of the mass inside would determine the equivalent heat capacity of a given reaction. If the products of combustion significantly affect the thermodynamic properties of the mass inside, calibration after combustion will be more meaningful.

If the thermal conductivity is known for a material, the measurement of area and temperature gradient could be used with Equation 8.10 to measure heat transfer. We are further removed from the basic definition, and we have more possibilities for errors. However, this method for measuring heat transfer is simpler to operate than the calorimeter. There are a variety of other methods for measuring heat transfer, but they are also derived from the basic concepts presented here. Extraneous inputs become more difficult to evaluate when the measurement systems are removed from the basic definition. In some instances simplicity of design, lower cost, and ease of operation justify the use of less accurate systems. If one is aware of the logical structure that links a measurement system to a physical parameter, it is possible to make some good approximations about the expected accuracy for a system. We note that the accuracy of measuring heat transfer using a calorimeter is dependent on the accuracy of measuring work, temperature, and pressure. (Mass is not required for many fluids.) Since heat is defined in terms of work and stored energy, it can be measured with less accuracy than either of these. Thermal conductivity is defined in terms of heat, but it is measured in terms of work. Still, it is less accurate than heat-transfer measurements, because it requires temperature gradients and area measurements in addition to heat measurements.

It is easy to see that we are getting further away (in the logical structure) from the basic standards. We expect to find less accuracy. This is equivalent to stating that there are more uncertainties. Thermal conductivity is relatively far removed from standards. We attempt to avoid defining other parameters in terms of this one. This is why we do not use the system of Figure 8.2 to measure length, area, or temperature.

8.4
Humidity
Content of Air

The properties of air are somewhat altered by the presence of water vapor. (The introduction of impurities into any system alters the system and the methods of evaluating these changes are the same as those presented for the present case.) It is important to be able to determine the moisture content of air in order to predict the thermodynamic properties of the air–water vapor mixture. Reference 2 gives a valuable source of thermodynamic properties for air–water vapor mixtures. These properties are available in both tabular and graphical form. A study of the moisture content of the air is called psychrometrics. The theory and basic concepts are presented in reference 3.

Humidity is defined to be the ratio of the actual partial pressure of water vapor to the saturation pressure of water vapor at the dry-bulb temperature. The term dry-bulb temperature is introduced in the study of psychrometrics to indicate a temperature determined when the sensor has no moisture on it. If moisture is maintained on the sensor surface by means of a saturated piece of cotton cloth, the temperature sensed by the measuring system will be lower than that of a dry-bulb thermometer. When water evaporates into the space around it, the energy of vaporization must be supplied by the thermal sensor, thus reducing the sensor temperature. The rate at which evaporation occurs depends on the relative humidity of the air in the space. When the relative humidity is 100 per cent, the evaporation rate is zero, and both the wet- and dry-bulb sensors measure the same temperature. This condition corresponds to the saturation line in reference 2.

Measurement systems must be devised to determine any combination of two of the following parameters in order to establish the thermodynamic state of the air at standard pressure: dry-bulb temperature dew point, wet-bulb temperature, humidity ratio (pounds-mass of water per pounds-mass of air), or relative humidity. The measurement of any two of these parameters can be used with the data of reference 2 to obtain all the others. The basic mathematical model for this phenomenon is very complex. Accuracies are best determined from tables of thermodynamic properties. Several methods of determining the thermodynamic state of moist air are shown in Figure 8.3. Here two of the list of above parameters are measured simultaneously for each system. Since dry-bulb temperature is easily measured, this will be assumed for each of the examples. In part (a) two thermometers are attached to a sling which can be rotated about the handle. The bulb of one is covered with a cloth wick and placed so that circular rotation will not allow any moisture to come in contact with the dry bulb. The rotation will cause the mercury column in the thermometer to be depressed. This must be corrected. A second commonly used wet bulb–dry bulb device employs a fan to force the air over the bulbs. These devices are simple to construct and easy to operate. Reference 1 discusses some of the capabilities of this device and some limitations.

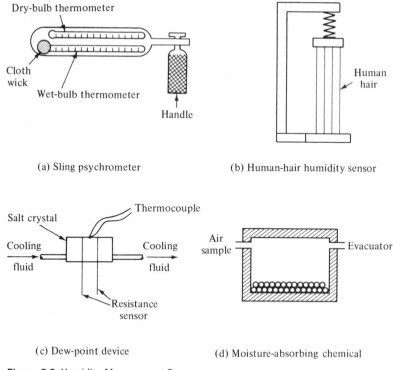

(a) Sling psychrometer

(b) Human-hair humidity sensor

(c) Dew-point device

(d) Moisture-absorbing chemical

Figure 8.3 Humidity Measurement Systems

The human-hair sensor of part (b) is frequently used in control systems. The length of human hair is very sensitive to the relative humidity in the air. This principle is used in some control devices to sense humidity ratio. The hair must be clean and must be exposed to the space. Calibration is obtained by one of the other methods discussed here.

A variety of dew point-type devices is available. The principle is to cool the air until the dew-point temperature is reached. Here moisture is condensed from the air to the surface at the dew-point temperature. Precise control of the surface is important. For part (c) the first appearance of humidity causes the salt crystal to become an electrical conductor. The resistance sensor records a finite resistance; the thermocouple measures the temperature at which this occurs. The device must then be heated above the dew-point temperature to be reused. The lower range is limited by the frost point (32°F), and the higher temperatures are limited by the crystal material.

If a sample of air is placed in the chamber of part (d) after the mass of the absorbing chemical has been determined, the change in mass of this chemical is dependent on the moisture absorbed. With a known volume, pressure, and temperature of gas, this provides the most accurate method of measuring air humidity. A typical absorbing chemical is phosphorous pentoxide. This is the basic calibration method and does require considerable experimental techniques.

Several other sensors for humidity employ the principles of variable thermal conductivity, spectroscopic methods, index of refraction, pressure or volume, thermal rise, mobility of ions, dielectric constant, critical flow, diffusion hydrometer, and chemical methods (reference 1). Reference 1 goes into some detail on the use of each of the above. After the required range and accuracy are known, a sensor can be selected that will satisfy the requirements. Variation in atmospheric pressure must be corrected using the data of reference 2.

8.5
Viscosity
Measurement

Isaac Newton first formulated the physical laws that defined the property viscosity. His hypothesis was that the force required to overcome the frictional effects of a fluid was proportional to the product of wall area and velocity gradient near the solid wall. The basic relationship for absolute viscosity is given by

$$F = \mu A \frac{du}{dy} \qquad (8.11)$$

where F = force
   μ = absolute viscosity
   A = wall area
   u = component of velocity parallel to the wall
   y = dimension perpendicular to the wall

All fluids which have a viscosity that can be represented by Equation 8.11 are called Newtonian fluids. These will be the only fluids considered here.

Two basic measurement systems are in common use for determining fluid viscosity. These are shown in Figure 8.4. The first is the concentric-cylinder experiment originally proposed by Newton. Here the force required to restrain the inner cylinder is measured by the spring. The cylinder area and the radial clearance are known from geometry. The velocity gradient is assumed to vary linearly between the velocity at the outside cylinder wall and the zero velocity at the inner cylinder. The distance over which this velocity is distributed is the radial clearance. Temperature, pressure, and fluid turbulence are extraneous inputs to this system. In more refined measurement systems, the entire system is submersed in a constant temperature

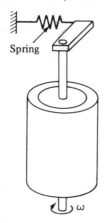

Spring

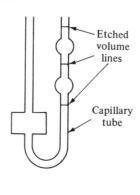

Etched
volume
lines

Capillary
tube

(a) Concentric cylinders   (b) Cannon-master viscometer

Figure 8.4 Viscosity Measurement
Systems

bath. Pressure actually introduces only a small error for most fluids. However, fluid turbulence can be a major problem, since it introduces other
transport phenomena into the measurement. References 1 and 4 are sources
of additional information.

The second measurement system of Figure 8.4 is used to measure the
kinematic viscosity through a capillary tube. The volume flow rate is determined by measuring the time required for the fluid to flow from the etched
mark between the bulbs to the etched mark below the bottom bulb.

Water is the standard fluid for viscosity. Its viscosity has been determined at 20°C using a capillary flow-type viscometer (reference 1). The
estimated accuracy is ± 0.000003 poise (1 poise $= 1$ dyne-s/cm²). The
value of viscosity of water at 20°C is 0.010019 poise. The defined standard
based on this experiment is 0.01002 poise. Water at 20°C can be used with
any viscometer to calibrate the device for the measurement of any other
fluid.

The basic equation for the capillary tube is

$$\frac{dP}{dL} = \frac{8\eta Q}{\pi r^2}$$

where $P =$ pressure
   $L =$ distance between two pressure taps
   $\eta =$ kinematic viscosity
   $Q =$ volume flow rate
   $r =$ capillary radius

The problem of determining the pressure gradient over the length of time of operation requires careful analysis. Reference 1 gives some of the details that must be considered in obtaining an accurate measure of viscosity. It also gives a bibliography of additional sources of information.

Since viscosity is related to fluid flow in a relatively complex manner, there are a variety of other methods used in the measurement of this property. A sphere may be dropped through a cylindrical container of a fluid and the time required for it to travel a given distance measured. This is very much like the capillary tube. The damping rate of ultrasonic waves has been proposed for measuring viscosity. Various modifications of the concentric cylinders have been used (two discs rotated with a fixed fluid thickness between them). The ASTM has publications on various methods of measuring viscosity.

With all defined parameters (work, heat, thermal conductivity, etc.), the extraneous inputs become more difficult to evaluate. It is not always possible to establish the complex interrelationships. The assumption that various properties have a linear effect on a parameter may not always be true. The NBS has looked at many of these problems and has provided guidelines in many problem areas. If doubt exists, they should be consulted. Methods of accurate measurement require much more complete analysis of the physical problem. To completely understand the type of analysis required with many of the measurement systems of this chapter, one would have to be an expert in each of the fields. Since this is asking too much of any one individual, an alternative solution has been adopted; if you are in doubt about extraneous inputs, accuracy, range, and dynamic response, consult the basic reference texts in the field or see the test codes. These are valuable sources of information, both for the measurement-system analysis and for the analysis of the engineering problem for which the measurement is required.

8.6
Chemical
Measurement
Systems

The amount of technical knowledge presently offered to the undergraduate scientist almost requires that the student specialize in some general area. This seems to be the only way to provide the basic physical background required in one area. The product is an individual with the ability to do basic research in one area and with very little capability in other areas. To offset this, it is presumed that the student can educate himself in a particular area that might be involved with his problem. One tends to forget the content of basic chemistry courses unless one is confronted with chemical-type problems. The measurement techniques associated with analytical chemistry (qualitative and quantitative) are seldom-used tools for many scientists. They cannot be treated in this book, but they are mentioned now to remind the reader that a variety of powerful experimental tools have been developed in this area.

In the analysis of chemically efficient power cycles, the chemical analysis of the products of combustion are important. The ever-expanding problem of pollution is also an area in which chemical analysis is being used. The techniques used in the Orsat analysis can be applied to a variety of gas-analysis problems. A typical Orsat apparatus is shown in Figure 8.5.

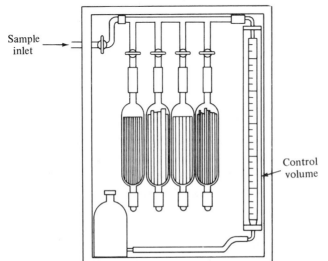

Sample inlet →

Control volume

Figure 8.5 Orsat Analysis System

A sample of gas is accepted into the volume meter, and the initial volume is determined. Then the gas is forced through a solution containing potassium hydroxide and water, where the carbon dioxide is absorbed. The volume is again measured, and then the gas is forced through the second absorbing medium (alkaline pyrogallol) and the oxygen is absorbed. The volume is measured once again to determine the volume of oxygen present in the original sample. The process can be repeated for other chemicals, such as carbon monoxide and sulfur dioxide, by using different absorbing solutions in the analyzer. Reference 5 gives the details of the operation of this system, and reference 6 states that the accuracy of measuring the volume content of any gas is, generally, less than 0.2 per cent of the volume measured. Certainly, more precise measurement techniques are possible; but the increase in accuracy causes an increase in analysis time. The Orsat analysis has been used very effectively for flue-gas analysis and internal-combustion-engine exhaust analysis. The ability to detect traces of chemicals in a sample space has grown considerably in recent years. The field of analytical chemistry has opened a great new potential for the measurement of chemical phenomena.

Another field that has been recently used in the measurement of engineering parameters is that of electrochemical kinetics (reference 7). Changes in chemical concentration, electrical fields, and pH factors have been used in the study of free-convective transport problems. The electrochemical field seems to be just opening up for general experimental use. The techniques have been known for some time, but applications are now being made at a faster rate. For example, suppose that a flat plate is placed in water. Salt is added to make it an electrical conductor, and an acid-indicating dye is added. A voltage is impressed between the lower edge of the plate and the metal bottom of the water container. At the surface of the plate, a locally acid condition can exist by electron exchange in the potential field. This local acid condition is detected by the indicator dye. One is able to observe the motion of a mass of fluid in a free convective field. Turbulent flow dissipates the local acid condition, and the fluid color returns to its initial condition. This tool has been used by several investigators for the analysis of free-convective heat transfer from a flat plate.

The mass spectrometer has become a powerful tool for detecting the presence of the chemicals present in a sample. The basic elements of a mass spectrometer are shown in Figure 8.6. An ionized sample is introduced into the chamber, or an ionizing system may be built with the spectrometer. The resulting ions are focused into a magnetic field, where their momentum is used to separate them into discreted bands. The location at which they strike the detector plate determines their mass and chemical composition. The chamber must be evacuated to pressures lower than 10^{-4} torr so that collisions with other molecues will be negligable. Reference 8 gives some

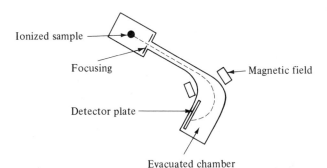

Ionized sample

Focusing

Magnetic field

Detector plate

Evacuated chamber

Figure 8.6 Mass Spectrometer

basic research that is being done in this field. It also lists references for anyone interested in doing work in this field. The basic physical laws, sensors, and accuracy are discussed in these references.

8.7
Acoustical
Measurement
Systems

The measurement of pressure waves propagated through air is important to the human environment factor. The basic sensors for this type of signal are shown in Figure 8.7. The condenser-type sensor uses a very thin metal diaphragm which deflects very easily under the action of the pressure waves.

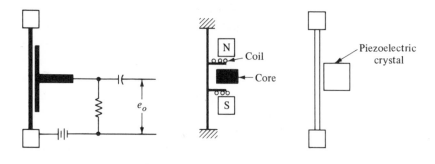

(a) Condenser microphone   (b) Moving-coil microphone   (c) Piezoelectric crystal

Figure 8.7 Pressure Wave Sensor

The output voltage is dependent on the distance between the diaphragm and the metal base. This type of microphone is used for reproducing most sound in the range of the human ear. The moving-coil type of sensor uses variable inductance for sensing the signal. The thin diaphragm is again the item that converts a pressure wave into a detectable form. This is also called the electrodynamic microphone. The frequency response of this system may tend to be somewhat limited by the mass attached to the basic diaphragm. The piezoelectric crystal has good dynamic response and most of the other capabilities required for a good sensor. It is also usually the most expensive.

All of the sensors mentioned here are basically for measuring the propagation of pressure waves in air. Similar types of sensors are used for underwater acoustics. A wide variety of detection technology is devoted to this type of sensor. Sensors are also important for determining the velocity of propagation in solids. Usually, crystal-type sensors are used in these latter applications.

The acoustical behavior of a mass has been used in the measurement of density, Young's modulus, shear modulus, Poisson's ratio, temperature, thermal conductivity, other transport properties, thickness, surface roughness, and many other parameter properties (reference 8). Certainly, the variety of parameters that are influenced by acoustical properties are extensive. Again we find that any effort at complete coverage would require too much time and effort. Therefore, only some very general comments on the acoustics of normal sound transmission will be mentioned.

In the sensing of sound in the range of the human ear, one is concerned with the dynamic nature of the phenomenon. If sound is to have its original meaning, the signal should not be distorted. This again involves dynamic calibration. For most sound-reproduction systems, the microphone may be calibrated using a pistonphone. This is a closed container fitted at one end with a piston, while the sensor to be tested is placed at the other end. The piston is driven at a range of known frequencies, and the sensor is calibrated. Many sensors can both sense a signal and be excited to produce a signal. These sensors may also be calibrated using the reciprocity relationship.

References 9, 10, and 11 give many of the extraneous inputs and methods of analysis that should be considered in the field of acoustics. The parameters that influence an acoustical wave are almost unlimited. The dust or humidity in the air, the temperature and pressure, mass transport, and any physical object that can reflect or distort a wave are only a few examples of the types of extraneous inputs that are possible. One must expect to devote a considerable amount of time to the field before one can reasonably expect to understand most of the variables that are important. Accuracy can be established for pure single-frequency tones, but the accuracy for more complex wave forms depends somewhat on the method of analysis. A starting point is the references suggested.

**8.8
Radioactive
Materials**

The alphatron used in the measurement of low pressures used a radioactive material to produce alpha particles. Several other measurement systems are now available which employ these techniques. The measurement of mass flow rate and velocity in a pipeline is possible using radioactive isotopes for tracers. Certain medical tests and treatments now use radioactive materials. The measurement of wear in machine parts has been made by these techniques.

The basic sensor for most of these radioactive materials is an ionization principle quite similar to the ionization gage in vacuum technology. However, here the basic cause for ionization of the material in a space is the radioactive particle or ray that is emitted when a radioactive molecule changes to a stable state. This product of radioactive decay will cause the mass of air inside a detector chamber to be partially ionized. The energy of the incoming particle or ray may cause several ionizations before the energy is absorbed or passes out of the chamber. The ionizations result is a flow of current that is higher than that associated with zero ionization. Therefore, the gage can detect the presence of high-energy particles and can provide some measure of their density in the space.

Radioactive materials must be handled carefully. They are not the sort of thing that one likes to have around the laboratory. However, they do represent a new technology and will undoubtedly find wider and more dramatic use in the future. A survey would not be complete without mentioning them. Reference 12 gives more specific experimental uses of radioactive materials. It also contains a short list of additional references on the subject.

**8.9
Conclusions**   This chapter pointed to several areas where different parameters must be measured. The method of calibrating measurement systems was generally very complex, and the NBS was suggested as a source of reference material.

The general technique for approaching any new measurement problem involves four steps.

1. Obtain the physical laws for a physical situation.
2. Write the governing equations when possible.
3. Select sensors for all variables required.
4. Analyze all measurements for accuracy.

If these procedures are followed, the measurement portion of the problem will generally not be a source of error. Errors at this point are usually the result of an incomplete understanding of parts 1 or 2 of this procedure.

A survey cannot cover the details or the basic techniques required to measure a parameter. It is important that much more research be done if the parameter to be measured is not well understood.

References

1. *Precision Measurement and Calibration, Heat and Mechanics*, NBS Handbook 77, vol. 2, U.S. Department of Commerce, Washington, D.C., 1961.

2. *ASHRAE Guide and Data Book, Fundamentals and Equipment*, ASHRAE, Inc., New York, 1963.

3. JORDAN, R. C., and PRIESTER, G. B., *Refrigeration and Air Conditioning*, 2nd ed., Prentice-Hall, Inc., Englewood Cliffs, N.J., 1956.

4. FULLER, D. D., *Theory and Practice of Lubrication for Engineers*, John Wiley & Sons, Inc., New York, 1965.

5. AMBROSIUS, E. E., FELLOWS, R. D., and BRICKMAN, A. D., *Mechanical Measurement and Instrumentation*, The Ronald Press Co., New York, 1966.

6. *Handbook of Measurement and Control*, "Instruments and Control Systems," Instruments Publishing Co., Pittsburgh, Pa., 1959.

7. VETTER, K. J., *Electrochemical Kinetics, Theoretical and Experimental*, Academic Press, New York, 1967.

8. MANNELLA, G. G. (ed.), *Aerospace Measurement Techniques*, NASA Sp-132, National Aeronautics and Space Administration, Washington, D.C., 1967.

9. PETERSON, A. P. G., and GROSS, E. E., JR., *Handbook of Noise Measurement*, General Radio Co., West Concord, Mass., 1967.

10. BROCH, J. T., *The Application of Brüell and Kjaer Measuring Systems to Acoustic Noise Measurement*, K. Larsen and Son, Denmark, 1967.

11. *Application of Brüell and Kjaer Equipment to Frequency Analysis Power Spectral Density Measurements*, K. Larsen and Son, Denmark, 1966.

12. BECKWITH, T. G., and BUCK, N. L., *Mechanical Measurements*, Addison-Wesley Publishing Co., Reading, Mass., 1969.

PROBLEMS

8-1. What is the basic definition of work? Based on this definition, what is the maximum accuracy with which work can be measured?

8-2. How does the local acceleration enter into the concept of work?

8-3. In terms of measurement, how can electrical power be equated to mechanical power?

8-4. Explain how the internal energy of air could be obtained experimentally. What is the zero internal-energy point? List each parameter that must be measured.

8-5. List all possible extraneous inputs to the solution of Problem 8-4.

8-6. Give an experimental apparatus for measuring thermal conductivity. Tell how its design helps to eliminate each of the major possible extraneous inputs.

8-7. What is the maximum accuracy with which thermal conductivity can be measured?

8-8. How can viscosity be determined from rotating concentric cylinders? What are the major extraneous inputs? How could a modified design be made to reduce these error sources?

8-9. How are kinematic viscosity and absolute viscosity related? Are the measurement systems equivalent?

8-10. What other gas-analysis techniques are possible in addition to the Orsat analysis? How does this relate to basic chemistry?

8-11. List several chemical changes that might possibly be used to measure a physical parameter.

8-12. How can two parameters (at atmospheric pressure) determine the humidity content of the air? Are all of the properties listed in Section 8.4?

8-13. Why is human hair used to measure relative humidity? How does the dew-point measurement differ from the wet-bulb temperature measurement?

8-14. Try to explain physically how acoustics could be used to measure Young's modulus, density, and temperature.

8-15. Would you be afraid to use a radioactive material in a measurement problem?

8-16. If your answer to Problem 8-15 was no, you may be a laboratory risk. Your answer should be yes until you know all the details and precautions necessary. Then you can make a definite response.

ELECTRO-OPTICAL
SYSTEMS

Optical measurement systems have, historically, made some of the greatest contributions to the physical sciences. The telescope was the basic instrument for collecting the data that resulted in a mathematical description of planetary motions. The Michelson–Morley experiment led to the formulation of the theory of relativity. The microscope has long been the basic measurement system in the biological sciences.

Recent developments in lasers, monochromers, and photoelectric sensors have resulted in more measurement capabilities. Optical memory devices are soon to be available. Holography is a blossoming new field for three-dimensional photography. High-speed cameras and variable-speed projectors are being used to analyze many phenomena. Thermal radiation has been explored more carefully because of the interest in the space industry. The polariscope is being used in stress-analysis problems for new transparent materials. The interferometer is being used in a variety of free-convection problems. The chemical analysis of gas is possible using colorimeters. The wavelength of light has become the standard for measuring length.

Present trends in optical systems are due to the major developments in electro-optical sensors and the laser. It is expected that many new optical systems will be developed and used in the immediate future. The most important aspects of these systems from a measurement point of view are that light waves are essentially frictionless, essentially inertialess, and easily amplified, they generally introduce only small extraneous inputs, and they are clean (no oil spots). These systems incorporate many of the most desirable features of a measurement system. With high accuracy and dynamic response they will provide better data in many physical fields. Therefore, we will discuss some of these systems to provide a better understanding of the basic principles of light.

9.2
The Telescope and Microscope

Both the telescope and the microscope are basic research measurement systems. They both have the same principle of operation, based on the focusing of light waves. Figure 9.1 gives the basic elements of a compound microscope. If a different set of lenses is used in the microscope body, the proper design of focal lengths and body size can be used to change the basic magnification of the object. The same type of design can be used to build a telescope. One of the most commonly used methods of calibration is to place an etched plate at the object and observe for length and distortion in the field of view.

The resolving power of the microscope is limited by the ability to distinguish two lines etched on a calibration plate. The smallest distance between distinguishable lines is the resolving power. This is partially dependent on the illumination that can be placed on the object. However, it is

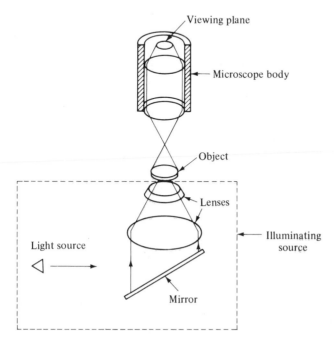

Figure 9.1 Basic Elements of a Microscope

basically dependent on the focusing and the lens in the microscope. The electron microscope uses a beam of electrons from an electron gun to illuminate the object. The electrons are reflected from the surface. Focusing and accelerating the reflected beam is accomplished by techniques similar to those used with the oscilloscope. The electrons must all have essentially the same initial velocity. The beam is projected on a phosphorous screen. The resolving power of the electron microscope is limited to the wavelength of the light associated with the energy of the electron. Theoretically, this is much shorter than the atomic spacing of the surface; but practically, resolution approximately equal to atomic spacing has been realized (reference 1).

The electron microscope has been used to observe atomic spacing of various surfaces. The view of grain structure lines should be observed with this technique. The analysis of lattice structure is also possible in crystals. A measure of surface roughness is possible with a focusing microscope. By focusing first on the peaks and then on the pits, the difference in distance between them can be measured. Work on energy sources and focusing techniques is underway, and better resolution will be accomplished.

The range of this type of instrument is limited by the basic optical design. However, a variety of instruments could be used to accomplish a desired range. These are essentially static measuring devices. If the output is photographed using high-speed cameras, the analysis of transient phenomena is possible. The limitation to dynamic measurement is essentially that of recording the output signal.

Monochromatic light sources (light of one wavelength) are used with the common microscope to observe some phenomena. Other analytical techniques are possible for this type of illumination. The microscope is a very important analytical tool, and many problems can be approached using this system. The basic principles of lenses and distortions can be found in several texts on optics. References at the end of this chapter can be consulted.

9.3
Monochrometers

Two very desirable characteristics for a light source are that it have single-wavelength (single-energy-level) properties and that it be well focused or collimated. A method of obtaining signle-wavelength light is possible from the basic laws of optics. The fundamental operating element of a monochrometer is the optical prism. In Figure 9.2, the prism is used to divide white light into its component parts. The white light strikes the prism as a plane wave. The index of refraction for any material is a function of the

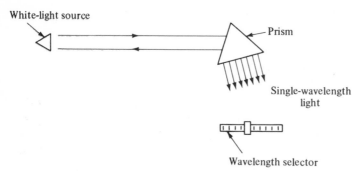

Figure 9.2 Basic Elements of a Mono-chrometer

wavelength of the light, and this dispersion effect is responsible for separating the white light into the single-wavelength positions of Figure 9.2. We could either rotate the prism or move the wavelength selector to obtain the output wavelength desired.

In many real monochrometers the light is actually reflected back through the prism to cause an additional spreading of the displaced waves. This is called a double-pass monochrometer. The prism material is selected for several reasons. Each material has a characteristic property for energy transmitted and for dispersion. The prism material is selected to provide the best possible wavelength selection and the maximum intensity of radiation. Usually, a collimator is provided to focus the output light. Many monochrometers also include a set of photoelectric sensors to measure the intensity of the separated waves.

A variety of basic optical instruments require a single-wavelength source. This is one function of the monochrometer. The system can be calibrated using an input signal that is from a single-wavelength source (mercury arc lamp, laser, etc). It can also be calibrated using a set of filters with the white-light source. In the case of single-wavelength sources, the position of the prism and the selector are calibrated for each wavelength generated by the source. When filters are used, only narrow wave bands of light are allowed to pass into the monochrometer. Again the prism and selector position can be calibrated for the known input signal.

The monochrometer was required for the measurement of thermal radiation properties of surfaces. It was determined that the analysis of thermal radiation properties for space application was essential. It was also found that the reflectance from a surface and the absorbance of a surface could be highly dependent on the wavelength of the incident wave. Therefore, the monochromatic properties of a surface had to be determined before a complete radiation heat-transfer analysis was possible. The basic measurement tool used in the evaluation of radiation properties of surfaces is the monochrometer.

The development of infrared detectors was also associated with the measurement of radiation properties. In all wavelengths, a sensor was needed to measure the intensity of the output wave. Ultraviolet radiation is a very short wavelength light wave (0.01 to $0.4\ \mu$), while infrared radiation is the long wavelength (0.7 to $10\ \mu$). Visible light varies from violet ($0.4\ \mu$) to red ($0.7\ \mu$). Detectors for wavelength outside the visible range require either photoelectric devices or thermal radiation collectors. Both are important in measurement technology.

The accuracy and range of the monochrometer depend on prism material and the precision with which the prism and selector can be positioned. The monochrometer is a static device. It can be constructed so that the spectrum of wavelengths of light can be continuously selected. A constant-wavelength drive is possible, so the wavelength can be varied linearly with time.

Spherical mirrors are incorporated to aid in collimating the output beam. Lenses and optical benches are also used to obtain a highly collimated beam. To obtain the desired features of a light source, both a monochrometer and a beam collimator may be used. These elements can provide a single-wavelength, highly collimated beam. One advantage of the monochrometer is that a range of wavelengths may be selected from this output. However, the output usually has a certain wave band of light and an associated collimator is required. The monochrometer may be used either as a source for light or as an experimental tool for measuring the wavelength content of an incoming light wave.

9.4
Lasers

A second type of light source is the laser. It has two very desirable features inherent in its operation. First, it is essentially a single-wavelength source. Secondly, it is very highly collimated. It has one disadvantage in that only one wavelength is generated from one particular laser. Therefore, the wavelength must be selected at the time the laser is chosen. The laser is not a measurement system. It does not have built-in sensors for determing the intensity of the output beam. However, it is one of the primary reasons why optical systems have enjoyed a revival of usage.

The laser beam has been concentrated for use in drilling small holes in metal. It has been used as a general heat source. It is a potential wavelength for transmitting voice communication. It has been used in measuring planetary distances. The laser has also been used in the measurement of local velocity in a clear-fluid flow field. In this mode of operation the beam is focused to a point in the fluid stream and the motion of the fluid causes a displacement of the output beam. The sensor and data analysis are relatively complex. It is claimed that the turbulent components of flow are measurable using this system. Lasers are used to weld printed-circuit elements. Scores of other uses are proposed, and laser technology is certainly in the developmental stage. Two other very important uses of the laser are in interferometry and holography. These applications will be discussed in this chapter.

9.5 Photoelectric Sensors

A variety of photoelectric sensors is now available for detecting a range of wavelengths. In the visible range, the photomultiplier tube is in common use. This operation allows a beam of photons to strike the cathode of the multiplier tube. The photons have enough energy to free electrons from the surface. The number of electrons freed is proportional to the photon energy and the number of photons striking the cathode. The electrons are attracted to the collector plate, and the resulting current flow is then a measure of the energy flux incident on the cathode. This particular sensor has frequently been used to open doors in supermarkets. It has also been used outside the visible light spectrum for burglar alarms.

Photodiodes operate on the same principle. A very wide range of solid-state devices is now available for use. A magazine devoted to advances in this area is *Electro-Optical Systems Design* (reference 2). Manufacturers and sensor data are presented in this publication. Many of the basic circuits and detectors are discussed in the technical portions of the magazine. Intermediate-circuit light modulators, laser sources, and detectors are all treated. The range and accuracies of various systems are presented. Fiber optics offer tremendous possibilities for intermediate modifying devices. The number of photosensors avilable in the new product section indicates the current growth in this field. The publication also presents complete systems for holographic and interferometer measurements.

One other type of sensor collects the energy from a beam of photons and stores the energy in a small mass of material. The resultant increase in temperature is measured using a thermocouple. This is also a measure of the energy of the incident beam of photons. The thermocouple sensor is a class of sensors based on the thermal radiation energy of a beam of photons. Thermistors and various thermistor-type sensors are also available in this area. The resultant increase in the temperature of the sensor is the measured output. Any change in an electrical property with temperature could be used in the same manner. Change in voltage output, current output, and resistance are common sensors. Each individual investigator has sensor types that he likes for each range. The range of incident wavelengths that must be measured places some constraints on the type of sensor required.

9.6
Interferometers

For two waves to be in a form that can interfere with each other to form standing waves, they must be coherent waves (i.e., they must come from the same source). If two waves from a single source travel different paths to reach a point, P, in space, they may reach the point in phase, in which case their amplitudes will add to produce a bright spot. They may arrive at this point $\frac{1}{2}$ wavelength out of phase, which will cause their amplitudes to cancel and produce a dark spot (reference 3). The mathematical model of light waves in this application is that they behave like electromagnetic waves. They have an amplitude and a frequency of oscillation.

The Michelson–Morley experiment mounted the interferometer of Figure 9.3 on a concrete mass and floated the mass on a mercury pool so that the

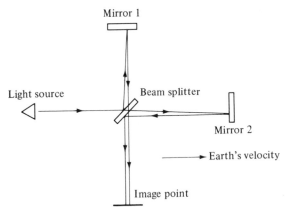

Figure 9.3 Michelson-Morely Interferometer

entire mass could be easily rotated. Mirrors were adjusted so that one line of beams was parallel to the earth's velocity and the other line was perpendicular to it. The time required to travel parallel to the earth's velocity would be slightly different than that required to travel perpendicular to the earth's field if the speed of light is dependent on the velocity of a reference frame. Then the entire interferometer was rotated through 90° to change the role of the beam paths. Light which was originally in phase at the image point should experience interference when the system was rotated. The measurement system was capable of sensing the change in path length due to the earth's velocity if it existed. The experiment determined that the speed of light did not depend on the earth's velocity. This was contrary to the scientific concepts at that time; Einstein was able to present a theory consistent with the experimental evidence.

The interferometer of Figure 9.3 could also be used to measure the displacement of either mirror 1 or mirror 2. In this case the change in only one path length would cause interference fringes at the image point. If a single-wavelength light source is used (a monochromater, a laser, a mercury arc lamp with filters, a sodium arc or other), the motion of $\frac{1}{4}$ wavelength of the mirror will increase the path length by $\frac{1}{2}$ wavelength. Then the two waves arriving at the image point will be $\frac{1}{2}$ wavelength out of phase, and destructive interference will occur.

The Mach–Zehnder interferometer is shown in Figure 9.4. It has been

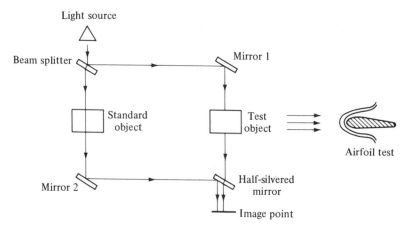

Figure 9.4 Mach–Zehnder Interferometer

used in a variety of two-dimensional fluid-flow and free-convective heat-transfer experiments. Light from a single-wavelength source is initially divided by a partially silvered plate. One path goes through the standard object chamber to mirror 2, then back to a second beam splitter. The other path is to mirror 1, through the test-object chamber and back to the half-silvered mirror. Both beams are joined here and travel to the image point. If the test object has a variation in the density in the fluid through which the light beam travels, the speed of travel depends on this density. Variation in density can result from the pressure variation around an airfoil (see Figure 9.4) or from variations in the temperature of the fluid around the test object (in free-convection heat transfer). In both of these cases the light passing adjacent to the test object may be adjusted to obtain constructive interference or a band of light. When the density changes linearly with position away from the test object, alternate destructive interference patterns are formed. The interferometer actually then measures the density variation in the fluid surrounding the test object. In the airfoil problem this essentially measures lines of constant pressure. In the convective heat-transfer problem this would measure isothermal lines.

A third type of interferometer used in measurement is the Fabry–Perot interferometer of Figure 9.5. This type of system is used in spectroscopic

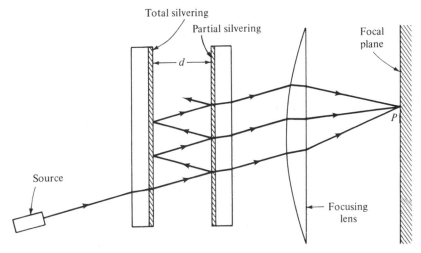

Figure 9.5 Fabry–Perot Interferometer

applications (references 3 and 4). Two partially silvered plates are placed parallel to each other at a distance d. At each reflection part of the incident energy is transmitted through the plate and is focused on the object plate by the intermediate lens. At point P the waves experience constructive interference and a line of light is present. If several light wavelengths are present in the light source, the output fringe patterns will be colored according to the wavelength present. This system is used to measure the wavelength of the incident light wave. It can be used in establishing the length of a meter in terms of wavelength of a monochromatic light source. It is perhaps the most versatile and easily adjusted interferometer. The resolution of these interferometers is basically that of $\frac{1}{2}$ wavelength of the light source for the first two systems. In the latter case, the presence of reflections makes it less amenable to simple analysis. However, multiple reflections make the basic requirements of monochromatic light sources more important. The resolution of general-measurement interferometers is dependent on the light source and the optics of the system. No other interferometers will be discussed here.

9.7
Holography

Holography was invented by Dennis Gabor in 1948, and it has found many applications in recent years. Light reflected from the object is illuminated on a photographic plate with no intermediate lens. If the light source, which is coherent, is also reflected from a mirror the resultant image on the photographic plate has a more realistic appearance than the normal photographic image. The resulting photograph is a hologram (reference 5).

The hologram records the interference patterns between the photographed object and the original light source. The photograph is placed in a beam of coherent light, where it produces a three-dimensional picture of the original object. The diffraction of crystals by X-rays can be viewed by holographic pictures. The results can be a study of the lattice structure of these crystals. Many new facets of this technique are still being discovered, and the limits have not been established.

One basic possibility for the holographic system is the measurement of surface roughness. The three-dimensional picture produced is certainly capable of sensing the depth of pits and the general surface contour. Usually, a laser is used for the light source.

The techniques of holography are still in their infancy, and future developments are certainly to be expected. A definitive measurement system is still to be developed for this new science. One would expect to see this technique developed to expand the capabilities of measurements. Holography is a static measuring system; its basic advantage is a three-dimensional picture of the object being considered. It is possible to view the object at various angles and to obtain useful information from the point of view selected. Reference 6 presents some advances in hologram technology.

9.8
Polariscopes

Photoelastic stress analysis has been achieved using a polarized light source of one wavelength. When polarized light passes through a medium where the lattice structure is not constant, the medium is optically active. In the unstressed state the index of refraction is constant in all directions, and the material is isotropic. However, when the material is stressed in a uniform condition the material acts like a uniaxial crystal and birefringent patterns result from the different rotation of the polarized light.

The stress patterns set up in a clear plastic substance may be viewed in the form of light and dark interference patterns. For some wavelengths of polarized light the colors may be blue and black, while for others it is white and dark. In any case it is possible to observe lines of constant stress and to obtain a stress analysis of the structure being stressed. This is a particularly useful measurement technique for regions where stress concentrations are present. Previous measurement methods were capable of measuring only surface stress, and mathematical models were proposed to predict the stress inside the body. With the present method it is possible to observe a two-dimensional state of stress in a body. If the stress is known at two opposite points in the body, it is possible to measure the stress from the optical fringes observed under a polariscope. Calibration can be accomplished by placing a body in simple bending, and the measurement of a stress concentration area can be accomplished from the original calibration.

9.9 Spectrometers and Colorimeters

The use of the spectroscopic lines to detect metallic components heated in a flame was first recognized by Kirchhoff and Bunsen in 1860 (reference 7). They observed that the wavelengths of light were independent of the flame temperature and that the number of lines observable increased with temperature. A measurement system capable of detecting the wavelength of emitted light is the monochrometer. When the monochrometer is used to analyze the spectrum of emitted light, it serves as a spectroscope. Spectrochemical analysis has become an accepted experimental tool for determining trace contents of a variety of metals. If the monochrometer is assumed to have optical focusing and detection equipment incorporated in it, the only additional requirement for analysis is the source.

The mechanism by which the atomic spectrum is excited will not be discussed in detail. However, the basic method is to energize the electrons to such a state that they will move to higher energy-level orbits. When they return to lower orbits, photons with a wavelength characteristic of the change in energy level are emitted. Each element has an atomic spectrum depending on the number of electrons in its orbits and the allowable excitation states. The flame from a Bunsen burner can excite some of the lower-energy-level transitions. Burning acetylene excites additional levels, and the electrical arc can be used to excite a much larger range of electron levels. Care must be taken to ensure that the energy source does not contribute to the spectrum observed.

Since each atom has its own characteristic atomic spectrum, it is possible to compare the data obtained from the monochrometer with the various spectra available. A match indicates the presence of that particular atom. Since the range of the spectra is from the ultraviolet to the infrared, it may be necessary to employ two or more prisms and detection systems. Spectroscopes also employ a grating in the measurement of the spectrum. This method will not be discussed here.

Calibration may be accomplished by filters, lasers, and known sample sources (reference 8). Possible error sources are also discussed in reference 8. The errors present are also those present in the measurement of the thermal radiation properties of materials.

There are several other factors that may be used in the analysis of a particular medium. The molecules that are illuminated by energy in all wavelengths will have characteristic absorption regions where all wavelengths in that region are absorbed. Other regions of the spectrum are allowed to pass through the liquid (or gas). The Beer–Lambert law is applied to this situation (reference 7):

$$\log_{10}\left(\frac{I_0}{I}\right) = kcl \tag{9.1}$$

where I_0 = the intensity at a given wavelength in air
I = the intensity passing through the medium
k = a constant for any wavelength
c = the concentration of the solute
l = the thickness of the absorbing medium

The law is valid only for monochromatic light. A variety of sensors are used at different wavelengths of the source illumination. The spectrophotometer is used to compare the intensity of the transmitted light with several standard samples. The colorimeter is used for only a few wavelengths.

The colorimeter uses two samples with different concentrations. The concentration of one is adjusted until the transmitted intensities are the same. Then Equation 9.1 can be used to obtain $kc_1l_1 = kc_2l_2$. Standard concentration cells must be maintained, and they must be periodically checked to ensure their value. The unknown concentration is either diluted or more solute is added to obtain the same transmission. Various monochromatic light sources are in common use, and the selection of a range must be included in the choice. Errors and calibrations are given in references 7 and 8. The detection may be in the form of photoelectric sensors. This method is a very convenient means of accurately measuring the associated intensities. The system can be used to determine the pH of a solute.

A modification of the colorimeter is used when the medium contains suspended opaque particles. Here the energy transmitted is different from that of colored molecules. The mechanism of absorption is not as clearly defined as in the previous case. However, careful calibration and a set of standards can be used to make essentially the same measurement using the same techniques.

One other method of measuring in an opaque medium is to detect the intensity of the light scattered at 90° to the incident illumination. This device is called a nephelometer. Another method (the fluorophotometer) measures the fluorescence of the molecules resulting from the incident illumination. Both methods require experimental calibration. The speed with which an analysis can be accomplished in an important factor in all of these systems. The accuracy, range, and calibration of the latter two devices is discussed in reference 7. Industrial and laboratory analysis of color, chemical traces, and concentration are accomplished using the systems of this section. In some cases they provide extremely accurate information. In other cases they are used to monitor a process, and a range of quality control is acceptable.

9.10 Photographic Records

One very powerful method for recording measured data is to make a photograph of the output information. Length calibration can be accomplished using a steel scale in the object plane. This method has been mentioned previously; high-speed cameras have also been mentioned.

The field of vision, focal length, and shutter speed all influence the quality of the picture taken. Camera optics are discussed in reference 9. The depth of field at the image plane is also a consideration in any camera application. The selection of camera optics capable of accomplishing a desired purpose is best made by consulting a good photographer. He can also suggest lighting and film for a particular application. In our society most families have at least one camera, and the reader may very well be acquainted with a variety of techniques in the field. An expenditure for a few rolls of film and some experimental photography may be in order before data collection is attempted.

The chemical aspects of the photographic process are presented in reference 10. It is generally not necessary to understand this process to take the picture. If the film is to be developed in the laboratory, a photographic supply house will provide a list of equipment required. Unless a large quantity of film is to be developed or a special developing process is required, commercial processing is recommended. Many quality processors are probably available locally, and several mail-order firms can be found.

9.11 Micro-densitometry

The image shown on a photographic plate may contain more information than is detected by the viewer. It contains degrees of illumination in addition to the shapes and general colors normally seen. Radiographic pictures transmitted from location to location must be critically viewed for contrast in order to make a valid reproduction of the signal. The microdensitometer measures the comparative transmittance of a light source through photographic film. It provides an intensity for each location on the film.

A slot allows a very small portion of the film to be illuminated from behind the film. A microscope and a detector are placed on opposite sides of the film to measure the amount of intensity transmitted. This information can be used to add to the contrast of a film or to provide additional information concerning the original image. It is particularly useful in the analysis of pictures taken of planetary objects or other space phenomena.

One additional use for this technique is to obtain spectral data on the emission of energy from a thermally radiating body. The radiating body is placed inside a sphere which is lined with an unexposed photographic film. The body is heated to a desired temperature; the radiation from this body exposes the film. The amount of exposure in any direction is recorded on the film. A microdensitometer can be used to measure the energy radiated in all directions. This is a very accurate method of determining emission properties of various thermal radiation materials. However, it is a very tedious and time-consuming method. The use of a coupled computer with the output sensor can be a great labor-saving device in this form of operation.

9.12 Refractometry

The measurement of the index of refraction of any medium can be used to determine the speed of light in that medium.

$$n = \frac{C_0}{C} \tag{9.2}$$

where n = the index of refraction of the medium
C = the speed of light in the medium
C_0 = the speed of light in a vacuum

When a beam of light passes from one medium to another, the beam is turned from its original path by an amount depending on the index of refraction of each medium. From the description of Figure 9.6 the relationship called Snell's law,

$$n_1 \sin \phi = n_2 \sin \theta \qquad (9.3)$$

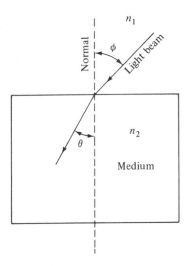

Figure 9.6 Index of Refraction

where n_1 and n_2 are the indexes of refraction in the corresponding media. The index of refraction in vacuum is 1, because of Equation 9.2. At one time air was used for the comparative index of refraction for all measurement systems. It was assumed to have an index of refraction equal to 1.0002925 $\pm 12 \times 10^{-7}$. Because of the many extraneous inputs into the use of air for the standard, reference 8 gives many of the parameters that must be observed and the associated corrections to the refractive index.

The refractive index of a medium is dependent on temperature, density, particle suspension, and many other parameters. Purity and concentration of solute can also affect this refractive index. Therefore, there are many measurement problems where this parameter could be used to obtain important information. Reference 7 gives several meters that are used to determine the refractive index of a liquid. The reference also gives the accuracy and range of these devices. Reference 8 gives more details of one of the instruments, along with the maximum accuracy obtainable. These references are recommended for those who wish to continue study in the area of index of refraction.

9.3
Conclusions
Electro-optical systems are important because they offer possibilities for precision in time and length. They may also have the highest response times of any sensor. The optical measurement of length is important in many fields. The microscope and the telescope are prime devices for measuring length.

The wave nature of light and the resulting interference patterns possible have made the field of interferometry especially important to measurements. It would be difficult to discuss all possible combinations of interference patterns possible. The ones mentioned here are very useful.

The monochrometer was presented for measuring the spectrographic intensity pattern for a variety of different physical situations.

Holography is a rapidly advancing field. More and more possible uses for this technique are being developed daily. When the laser was developed, the field of holography was actually the most influenced field. A study of optical systems is urged so that one is able to keep abreast of the fastest growing area of measurement.

References

1. HAINE, M. E., and COSSLET, V. E., *The Electron Microscope*, E. and F. N. Spon, Ltd., London, 1961.

2. *Electro-Optical Systems Design*, Milton S. Kiver Publications, Inc., Chicago, Ill., 1970.

3. TOLANSKY, S., *An Introduction to Interferometry*, John Wiley & Sons, Inc., New York, 1955.

4. ROSSI, B., *Optics*, Addison-Wesley Publishing Co., Inc., Reading, Mass., 1957.

5. MANNELLA, G. G. (ed)., *Aerospace Measurement Techniques*, NASA SP-132, National Aeronautics and Space Administration, Washington, D.C., 1967.

6. *Proceedings of the Symposium on Modern Optics*, New York, 1967, Microwave Research Institute Symposia Series, vol. 17, Polytechnic Press, Polytechnic Institute of Brooklyn, Brooklyn, N.Y., 1967.

7. GIBB, T. R. P., JR., *Optical Methods of Chemical Analysis*, McGraw-Hill Book Co., Inc., New York, 1942.

8. *Precision Measurement and Calibration-Optics, Metrology, and Radiation*, NBS Handbook 77, vol. 3, U.S. Department of Commerce, Washington, D.C., 1961.

9. BORN, M., and WOLF, E., *Principles of Optics*, Pergamon Press, Elmsford, New York, 1959.

10. BOWEN, E. J., *The Chemical Aspects of Light*, The Clarendon Press, Oxford, 1946.

PROBLEMS

9-1. Draw the optical paths for a given lens to show that it can magnify an object. What is the basic curvature of the lens for magnification?

9-2. How can a microscope be calibrated?

9-3. What is a monochrometer? How does it work?

9-4. If white light is the input to a monochrometer, what is the output?

9-5. Give three methods by which a monochrometer can be calibrated.

9-6. What is a laser? How does it work? What are its main uses?

9-7. What are the capabilities and limitations of the laser in comparison to the monochrometer?

9-8. Tell how each interferometer works. How could each be calibrated? What can each measure?

9-9. Can you think of any other physical problem where an interferometer could be used?

9-10. What are the basic differences between each of the interferometers presented here?

9-11. How can a spectrometer be used in chemical analysis? How does it compare with a monochrometer?

9-12. Name two optical measurement systems not discussed in this book. Tell how they work.

9-13. How could lasers be used to determine liquid level? How could an interferometer be used to accomplish the same purpose?

9-14. What is the most accurate length measurement possible using an interferometer with a monochromatic light source of 2 μ.

9-15. How could light be used to measure density?

9-16. A circular disc is placed on a human hair in a vacuum chamber. It is illuminated from a known source. Tell how it could be used to measure mass and force.

I

REPORT WRITING

Perhaps the greatest obstacle that presently confronts the engineer is that of communication. The greatest discovery of the century is valueless if the discoverer is unable to present his discovery to others. Therefore, all engineers must be able to present their work in a form that is complete and easily understandable. One absolute requirement for the completion of an education is a demonstrated ability to write a report.

You are personally responsible for fulfilling the requirements of completeness, understanding, and conciseness. However, the following general form of a report has proved adequate in many engineering problems. If you have doubts, the reference books listed at the end of this appendix will prove useful.

AI.1
Contents of a
Report

1. Title Page

    The report title should prepare the reader for the contents of the report. It should be as concise as possible. The name of the report author(s) should appear below the report title.

2. Abstract

    An abstract is a very critical part of a report. The purpose of this section is to present the problem and the conclusions resulting from the investigation of this problem. This should be accomplished using a very minimum of words. (Check articles in TASME journals.)

    Many people read only the abstract to determine whether or not the report should be read in detail. Do not slight the work you have done by presenting an inadequate abstract.

3. Table of Contents

4. List of Symbols

5. Introduction and Literature Review

Other people reading your report may not be familiar with previous work done in this field. It is necessary to introduce them to the problem area and to acquaint them with the work that has preceded the present investigation. This section should also explain why the present investigation was undertaken.

6. Analytical Approach

Every experiment you perform will be an attempt to associate a measured phenomenon with an analytical model described by some physical analysis. This analysis usually takes the form of a mathematical model of the phenomenon. It is possible in the initial phases of this course that a mathematical model has already been postulated and is available in several texts. However, this fact does not relieve you of the responsibility of presenting the presently accepted model. Liberal use of acceptable references is expected in this section.

7. Experimental Program

This section should explain the experimental apparatus, the measurements taken, the accuracy of these measurements, and extraneous inputs that might influence the experiment. The techniques used in taking measurements should be explained, and any errors from the assumed model should be discussed.

8. Results

The results of the investigation should be presented in either graphical or tabular form. Present the results of your experiment so that what you have accomplished is immediately apparent. It is also important to present (on the same graph or in the same table) the results expected from the analytical model and the results found experimentally by other similar experiments.

9. Conclusions

This section includes comparisons of theoretical and experimental results (usually in per cent difference). Any positive statements concerning the results of this experiment should be made (i.e., (a) this experiment correctly represents the physical conditions of the problem; (b) the data collected may be used with an expected error less than —— per cent). This is another section of the report that may be read by many engineers if the abstract was interesting.

10. Appendix

All data collected in the laboratory should be included here, along with sample calculations and any procedures or methods that were too elaborate to be included under experimental program. Mathematical developments that were mentioned in the analytical approach but were too long to be completed there may be placed in the appendix. Anything that does not add to the presentation in a section but that should be included somewhere in the report can be incorporated in the appendix.

AI.2
Concluding
Remarks

This book was not designed to teach report writing, but all engineers must know how to present results properly. When you arrive in industry, you may be known to your superiors only by the reports you submit. Therefore, you must become an accomplished report writer even if it requires a self-taught course. The challenge must be met by you. The following references are suggested for your guidance.

References

1. CROUCH, W. G., and ZETLER, R. H., *A Guide to Technical Writing*, 2nd ed., The Ronald Press Co., New York, 1948.

2. HOLSCHER, H. H., *How to Organize and Write a Technical-Report*, Littlefield, Adams & Co., 1965; KEREKES, F., and WINFREY, R., *Report Preparation*, 2nd ed., Iowa State College Press, Ames, Iowa, 1951.

3. SOUTHER, J. W., *Technical Report Writing*, John Wiley & Sons, Inc., New York, 1957.

4. ULMAN, J. N., JR., and GOULD, J. R., *Technical Reporting*, rev. ed., Holt, Rinehart and Winston, New York, 1959.

5. ZALL, P. M., *Elements of Technical Report Writing*, Harper and Row, Publishers, New York, 1962.

II

DIMENSIONAL ANALYSIS

The introduction to a dimensional system in Chapter 6 clearly demonstrates that consistency in any dimensional system depends on the definition of at least three dimensions. It is mentioned that length, time, and mass are three basic dimensions for measurement. It is also pointed out that temperature, voltage, and current (although basically defined by these three selected dimensions) have a complex interrelationship with the three basic dimensions. Therefore, temperature and current are also frequently defined dimensions consistent with the three selected dimensions. These two dimensions are often considered to be independent in a dimensional analysis.

Dimensional analysis, then, is a method of approaching a physical problem employing the fact that all dimensions are related to the basic dimensions in some manner. This essentially says that the particular parameters, incorporated in the mathematical model of physical phenomenon, are interrelated with the dimensional system selected. The mathematical model must be dimensionally consistent if it is to represent the physical phenomenon properly. In the consideration of a physical problem, the functional relationship between the important parameters can be partially determined by the requirement that they be dimensionally consistent. This last point is the foundation on which the principles of dimensional analysis are based.

There are two basic methods available for use in dimensional analysis of a physical problem. First, a proposed mathematical model may be available which defines the relationship between the various parameters involved in the problem. For this case let us look at the problem in which a steel shaft is loaded in tension. The mathematical model of the resultant change in shaft length is related by the equation

$$\frac{F}{A} = E\frac{\Delta L}{L} \tag{AII.1}$$

where $F =$ applied force
   $A =$ cross-sectional area of shaft
   $E =$ modulus of elasticity
   $\Delta L =$ change in shaft length
   $L =$ original shaft length

Now we express all of these parameters in terms of the basic dimensions: time, t; length, L; and mass, M. Then,

$$F = \frac{ML}{t^2}$$

$$A = L^2$$

$$\Delta L \text{ and } L = L$$

and

$$E = \frac{M}{Lt^2}$$

The term $\Delta L/L$ is already dimensionless, since the mathematical laws for addition, subtraction, division, and multiplication are also valid for dimensional units. Therefore,

$$\frac{\Delta L}{L} = \frac{\text{length}}{\text{length}} = 1$$

is a dimensionless group of two parameters and is called strain, ϵ. Also,

$$\frac{F}{A} = \left(\frac{ML}{t^2}\right)\left(\frac{1}{L^2}\right) = \frac{M}{Lt^2} = \text{stress} = S$$

Both sides of the mathematical model can be made dimensionless by dividing each side by E. Equation AII.1 becomes

$$\frac{S}{E} = \epsilon$$

The dimensionless groups important in analyzing the loading of a shaft are strain, ϵ, and S/E. Example 2.1 discussed this particular relationship.

   This first method is the result of the nondimensionalization of a physical law. To realize fully the capabilities of this method in an experimental analysis, the basic physics of the problem must be understood. This method introduces the possibility of similarity modeling. This is a powerful experimental tool which allows the testing of an experimental model to provide information about a similar physical situation. This principle is used extensively in strength of materials, fluid mechanics, aerodynamics, and transport problems. In the above example, one could test a 1-in.-diameter shaft to obtain information about a similarly loaded 6-in.-diameter shaft. One additional example of this method will be presented to outline the capabilities.

The one-dimensional momentum equation for an incompressible fluid with zero body forces is given by Equation AII.2,

$$\rho u \frac{du}{dx} = -\frac{dp}{dx} + \mu \frac{d^2u}{dx^2} \qquad\qquad\qquad \text{(AII.2)}$$

where u = fluid velocity
          x = linear distance
          p = fluid pressure
          ρ = fluid density
          μ = fluid viscosity

This equation can be nondimensionalized by first defining the dimensionless ratios

$$v = \frac{u}{u'} \quad \bar{x} = \frac{x}{x'} \quad \text{and} \quad \bar{p} = \frac{p}{p'}$$

Equation AII.2 becomes

$$\left(\frac{\rho u'^2}{x'}\right)(v)\frac{dv}{d\bar{x}} = -\left(\frac{p'}{x'}\right)\frac{d\bar{p}}{d\bar{x}} + \left(\frac{\mu u'}{x'^2}\right)\frac{d^2v}{d\bar{x}^2} \qquad\qquad \text{(AII.3)}$$

The only terms having dimensions in Equation AII.3 are those in parentheses. If we divide both sides of the equation by any one of these terms (let us divide by $\rho u'^2/x'$), the result is

$$(v)\frac{dv}{d\bar{x}} = -\left(\frac{p'}{\rho u'^2}\right)\frac{d\bar{p}}{d\bar{x}} + \left(\frac{\mu}{\rho u' x'}\right)\frac{d^2v}{d\bar{x}^2} \qquad\qquad \text{(AII.4)}$$

Now, since the left-hand side of Equation AII.4 is dimensionless, the terms in parentheses on the right-hand side must also be dimensionless. We can check this:

$$p' = \frac{M}{Lt^2} \qquad \rho = \frac{M}{L^3} \qquad u' = \frac{l}{t}$$

and

$$\frac{p'}{\rho u'^2} = \left(\frac{M}{Lt^2}\right)\left(\frac{L^3}{M}\right)\left(\frac{t^2}{L^2}\right) = 1$$

This term is the ratio of the pressure forces to the inertia forces for this problem. The second term, $\mu/\rho u'x'$, is the ratio of the viscous forces to the inertia forces and is divided by Reynolds' number, where Reynolds' number is

$$\rho\frac{u'x'}{\mu} = \frac{(M/L^3)(l/t)(L)}{(M/Lt)} = 1$$

Two different physical situations will be similar if the numerical values of these two dimensionless groups are identical. This allows us to test what occurs in water and to predict what will happen in alcohol. One might also predict what would happen in air if the air flow were incompressible. The reader is referred to the references for more details about similarity modeling. Similarity is also used in analog models where electrical circuits model mechanical spring–mass systems.

The second method available for dimensional analysis is used when the mathematical model is unknown. This method is used to assist in experimentally determining the mathematical model. In this second method, one must first list every physical parameter that influences the problem under consideration. If any parameters are overlooked, the method will fail to indicate any dependency on this parameter. The equations may be dimensionally inconsistent in this case. However, this method will provide a method of determining the dimensionless groups that are important to the physical problem. This is accomplished as a result of the step-by-step method proposed by Ipsen (reference 1) or by the Buckingham π theorem. The Buckingham π theorem says that the number of dimensionless groups obtainable from a given set of parameters is equal to the number of parameters minus the number of basic dimensions required to define the parameters. One problem with this theorem is that all of the dimensionless groups may not be independent of each other. This problem is avoided by the Ipsen approach. It is necessary to know something about a physical problem before one can select the parameters that are important to the problem solution. This knowledge must be gained by the reader in his particular field. This appendix is concerned only with the methods for obtaining the dimensionless groups.

In the step-by-step approach, some additional information on the independence of the dimensionless groups is available. This method will be demonstrated first using an example. Then the Buckingham π theorem will be used.

The dynamic response of a thermocouple to a transient gas temperature is to be analyzed. From a physical point of view the thermocouple junction must be heated up by the convective-heat transfer to the junction. This involves the heat-transfer rate to the junction from the gas, $q = hA_j(T_g - T_t)$. It also involves the energy stored at the junction, $q = mc\, dT/dt$, and the energy conducted down the thermocouple wires, $q = KA_w\, dT/dx$. For the present analysis, the parameters of importance are the convective-film heat-transfer coefficient, $h(LF/tL^2T)$, the area of the junction, $A_j(L^2)$, the gas temperature, T_g, the thermocouple junction temperature, T_t, the mass of the junction, m, the specific heat of the junction, $c(LF/MT)$, and the change in junction temperature with respect to time, $(T_t - T_0)/\Delta t$. The conduction down the wire is neglected for the present analysis. Then we express these parameters in mathematical form,

$$T_T - T_0 = f(h, A_j, m, c, t, T_g - T_t)$$

where T_0 is the initial thermocouple temperature. We now proceed to eliminate each of the basic dimensions, length, mass, time, and temperature, from the expression. Temperature appears in the terms h and c, along with the two temperature difference terms (remember that $F = ML/t^2$). There is some choice available and some engineering experience is valuable in making the best selection. Since $T_g - T_t = T_g - T_0 - (T_t - T_0)$, the last parameter in the function can be replaced with $T_g - T_0$. Division by this term gives

$$\frac{T_t - T_0}{T_g - T_0} = f(h, A_j, m, c, t)$$

Since h and c still contain temperature, they are also divided to obtain

$$\frac{T_t - T_0}{T_g - T_0} = f\left(\frac{h}{c}, A, m, t\right)$$

Now, h/c has dimensions M/tL^2. We can eliminate mass by dividing this ratio and the term m by mass to obtain

$$\frac{T_t - T_0}{T_g - T_0} = f\left(\frac{h}{cm}, A_j, t\right)$$

Since time appears in only two terms, this dimension can be eliminated by collecting time with this ratio:

$$\frac{T_t - T_0}{T_g - T_0} = f\left(\frac{ht}{mc}, A_j\right)$$

The same applies for length squared:

$$\frac{T_t - T_0}{T_g - T_0} = f\left(\frac{hA_{jt}}{mc}\right)$$

Now all groups are dimensionless, and it should be possible to obtain the functional relationship between these parameters experimentally.

Now let us solve the same problem using the Buckingham π theroem, but we will include the conduction term. The functional relationship involves

$$f(T_t - T_0, h, A_j, m, c, t, T_g - T_0, K, x, A_w) = 0$$

One parameter must be chosen to be the dependent variable. We might choose $T_t - T_0$. The Buckingham π theorem states that there are $10 - 4$ (number of parameters minus number of basic dimensions) $= 6$ dimensionless groups. Since $T_t - T_0$ was selected to be the dependent parameter, it appears in only one dimensionless group. One obvious method of selecting a proper group is to include only the term $T_g - T_0$. Then

$$\text{group } 1 = \frac{T_t - T_0}{T_g - T_0}$$

One mathematical method for obtaining this reduction is to form a table similar to Table AII.1. Since h has dimensions M/t^3T, the exponents

TABLE AII.1

	K_1 $(T_g - T_0)$	K_2 h	K_3 A_j	K_4 c	K_5 K	K_6 x	K_7 M	K_8 t	K_9 A_w	K_{10} $(T_t - T_0)$
M	0	1	0	0	1	0	1	0	0	0
L	0	0	2	2	1	1	0	0	2	0
T	1	-1	0	-1	-1	0	0	0	0	1
t	0	-3	0	-2	-3	0	0	1	0	0

of these dimensions are placed in the table. The rank of the 4×10 matrix of the table is obviously 4. If one writes the algebraic equations involved with dimensionally consistent equations, the following equations result:

$$K_2 + K_5 + K_7 = 0$$
$$2K_3 + 2K_4 + K_5 + K_6 + 2K_8 = 0$$

$$K_1 - K_2 - K_4 - K_5 + K_{10} = 0$$
$$-3K_2 - 2K_4 - 3K_5 + K_8 = 0$$

Here are four equations with 10 unknowns, We can solve for four unknowns in terms of six exponents. Here $K_1, K_2, \ldots, K_6$ are selected to be independent exponents. Then

$$K_7 = -K_2 - K_5$$
$$2K_9 = -2K_3 - 2K_4 - K_5 - K_6$$
$$K_{10} = -K_1 + K_2 + K_4 + K_5$$
$$K_8 = 3K_2 + 2K_4 + 3K_5$$

Now we are at liberty to select any values for the six independent exponents. Six independent dimensionless groups will result when the six different choices for these exponents are independent of each other. For our example the table of choices are given in Table AII.2. It is seen that groups 1, 2,

TABLE AII.2

Group No.	K_1	K_2	K_3	K_4	K_5	K_6
1	1	0	0	0	0	0
2	0	0	1	0	0	0
3	0	0	1	0	0	-2
4	0	1	1	-1	0	0
5	0	0	0	-1	1	-1
6A	0	1	1	0	-1	1
6B	0	0	0	1	0	0

3, 4, 5, and 6A are not independent, since the determinant of their matrix of coefficients is zero (row 4 — row 5 — row 6 = the zero row, 0 0 0 0 0 0). Row 6 B was selected to give six independent dimensional groups. Now the determinant of the matrix of coefficients of rows 1, 2, 3, 4, 5, and 6B is not zero, and the choice of these exponents gives independent dimensional groups based on the six independent parameters selected.

The dimensionless groups are usually selected based on the experience of the user. This is not always the most logical approach, but it does very rapidly provide for the selection of several prominent dimensionless groups. This method is used in the following discussion.

A second dimensionless group is associated with geometry

$$\text{group 2} = \frac{A_j}{A_w}$$

or

$$\text{group 3} = \frac{A_w}{x^2}$$

The next group involves the convection and stored-energy terms: h, A, m, c, t. When the dimensionless groups are not obvious, an algebraic equation can be formed to establish the dimensionless group in the following manner:

$$h^a, A_j^b, m^c, c^d, t^e = (M^0, L^0, T^0, t^0)$$

Now we consider the dimension of each parameter:

$$\left(\frac{L^2 M}{t^3 L^2 T}\right)^a (L)^{2b} (M)^c \left(\frac{L^2 M}{MTt^2}\right)^d (t)^e = (M^0, L^0, T^0, t^0)$$

Collecting exponents of each term from this gives

$$(M)^{a+c}(L)^{2d+2b}(T)^{-a-d}(t)^{e-3a-2d} = M^0, L^0, T^0, t^0$$

Equating exponents on each side of the equation gives

$$a + c = 0 \quad (M)$$
$$2d + 2b = 0 \quad (L)$$
$$-a - d = 0 \quad (T)$$
$$e - 3a - 2d = 0 \quad (t)$$

These are four equations in five unknowns, requiring a specified value to be arbitrarily selected for one exponent. We select $a = 1$. Then

$$c = -1, \quad d = -1, \quad b = 1, \quad \text{and} \quad e = 1$$

The dimensionless group is

$$\text{group 4} = \frac{hA_j t}{mc}$$

A fifth group involves the energy conducted down the wire and the energy stored:

$$K^a A_w^b x^c m^d c^e t^f = M^0 L^0 T^0 t^0$$

This gives

$$\left(\frac{L^2M}{t^3LT}\right)^a(L)^{2b}(L)^c(M)^d\left(\frac{L^2M}{MTt^2}\right)^e(t)^f = M^0L^0T^0t^0$$

The resulting algebraic equations are

$$a + 2b + c + 2e = 0 \quad (L)$$
$$a + d = 0 \quad (M)$$
$$a + e = 0 \quad (T)$$
$$-3a - 2e + f = 0 \quad (t)$$

These are four equations in six unknowns, requiring the selection of two parameters. Let $a = 1$ and $b = 1$. Then $e = -1$, $d = -1$, $f = 1$, and $c = -1$. The dimensionless group is

$$\text{group } 5 = \frac{KA_w t}{xmc}$$

A sixth dimensionless group involves a balance of the energy in by convection and the energy out by conduction:

$$h^a, A_j^b, K^c, x^d, A_w^e = M^0L^0T^0t^0$$

This gives

$$\left(\frac{M}{t^3T}\right)^a(L)^{2b}\left(\frac{LM}{t^3T}\right)^c(L)^d(L)^{2e} = M^0L^0T^0t^0$$

The algebraic equations are

$$2b + c + d + 2e = 0 \quad (L)$$
$$a + c = 0 \quad (M)$$
$$-a - c = 0 \quad (T)$$
$$-3a - 3c = 0 \quad (t)$$

These are actually only two independent equations with five unknowns. Therefore, we can select three exponents. Let $a = 1$, $b = 1$, and $d = 1$. Then $c = -1$, and $e = -1$. The dimensionless group is

$$\text{group } 6A = \frac{hA_j x}{kA_w}$$

Since A_j/A_w is already a dimensionless group, this factor may be removed to give

$$\text{group 6A} = \frac{hx}{K}$$

However, it may be seen that group 4 divided by group 5 gives group 6A. This group is not independent of the previously determined groups. If a table similar to Table AII.2 is constructed at this point, it is easy to determine that the last independent dimensionless group involves the energy stored in a given time.

$$C^a, M^b, t^c, A_w^d(T_g - T_0)^e = M^0L^0T^0t^0$$

The algebraic equations are

$$b = 0 \quad (M)$$
$$2a + 2d = 0 \quad (L)$$
$$-a + e = 0 \quad (T)$$
$$-2a + c = 0 \quad (t)$$

We select $a = 1$. Then $b = 0$, $c = 2$, $d = -1$, $e = 1$, and the dimensionless group is

$$\text{group 6B} = \frac{Ct^2(T_g - T_0)}{A_w}$$

In this particular problem the area involved may not be the wire cross-sectional area, and the temperature difference may not be the gas temperature minus the initial thermocouple temperature. These problems must be evaluated based on the physics of the situation. However, some characteristic area and temperature difference are involved with this last dimensionless group.

One advantage of the tabular method of finding the dimensionless groups is that one is able to select the exponents in such a way that the control parameters appear in only one dimensionless group. This offers tremendous advantages in the final analysis of the problem. It also simplifies data collection, in that fewer points are normally required. In experimental analysis this feature is a powerful tool. For the above problem the convective heat-transfer coefficient and gas temperature would be the basic control parameters and the thermocouple temperature would be the measured output. Variation in other parameters would require material and geometry changes.

Before an experimental program is started, some type of dimensional analysis should be made. Either the basic mathematical model is known and the dimensionless groups are easily determined, or the mathematical model is unknown and the step-by-step or tabular method should be used. In any case, knowing the dimensionless groups involved in a problem can simplify both the analysis and the experimental control of the experimental program.

References

1. IPSEN, D. C., *Units, Dimensions, and Dimensionless Numbers*, McGraw-Hill Book Co., Inc., New York, 1960.

2. LANGHAAR, H. L., *Dimensional Analysis and Theory of Models*, John Wiley & Sons, Inc., New York, 1951.

III
CALIBRATION
FUNDAMENTALS

The object of calibrating a measurement system is to prove that the measurement obtained can adequately represent the input signal being measured. This proof is in the form of logic that the measurement of a known input signal by the measurement system gives a particular output signal. When an unknown signal produces the same output under the same conditions, the value of the unknown variable must be equal to the known value that produced this output in the calibration procedure. (Values of a variable producing the same output on a measurement system under the same conditions must be equal to each other.) When static calibration is performed, a plot of system output versus known input is called a calibration curve. The output of the system measuring an unknown signal can then be used with this calibration curve to establish the value of the unknown input. The calibration curve could be eliminated if the mathematical relationship between input and output information could be determined.

The most essential part of evaluating measurement systems is the method of establishing the relationship between the measured output and the associated input. Any calibration process must start by having a known input signal. This signal is called a standard. Standards are defined in Chapter 6 for the basic dimensions of length, mass, time, temperature, and current. All other dimensions are based on these. Primary standards are seldom available in the laboratory. Secondary standards are available from the NBS for use in calibrating most measurement systems. These must be selected with care if they are to provide the range and capabilities needed.

Suggested needs for a standard laboratory are given in reference 1. Also see the NBS volumes referenced in Chapter 6.

AIII.1
Static
Calibration

Calibration is considered in two parts: static and dynamic. Static calibration is accomplished by applying a range of standard inputs that extends beyond the range for which the instrument is to be used. The input signals are discrete and constant, but different values are required. In dynamic calibration the input signal must also be known, but it is a continuous function of time (this is opposed to the discrete nature of the static input signal). The basic difference between these two types of calibration can best be seen from the mathematical representation of a general measurement system:

$$(A_n D^n + A_{n-1} D^{n-1} + \cdots + A_1 D + A_0)I_o$$
$$= B_m D^m + B_{m-1} D^{m-1} + \cdots + B_1 D + B_0)I_i \qquad \text{(AIII.1)}$$

where the A's and B's are constants, D^k is the operator derivative of order k, I_o is the information out of the measurement system, and I_i is the input information. For static calibration the input signal becomes $B_0 I_i$ and the output (steady-state) signal is $A_0 I_o$. In the solution of the differential Equation AIII.1, the left-hand side is set equal to zero to obtain the homogeneous solution. This solution characterizes the measurement system for any input signal. It is the transient response of the system. However, when the input is a static signal, the transient terms will tend toward zero if the system is to be an effective measurement device. For static calibration, all derivatives are set equal to zero and the gain is given by

$$\frac{I_o}{I_i} = \frac{B_0}{A_0} = K \qquad \text{(AIII.2)}$$

Static calibration is used to determine the constant K. For dynamic calibration the B's are all known, since the input signal must be known for calibration to be possible. Dynamic calibration is used to determine all of the other A's in Equation AIII.1 (A_0 is known from the static calibration).

In static calibration, standard input signals should be known with an accuracy approximately 10 times that of the system being calibrated. (Standards that are only two times as accurate as the system being calibrated have been used effectively, but much more careful techniques must be applied to eliminate extraneous inputs.) If the system being measured is accurate ± 0.1 unit, the standard should be accurate ± 0.01 unit if it is to be 10 times as accurate as the system.

If readings are carefully taken, bias-type errors can be evaluated. If the standard deviation of the output signal is small (see Chapter 5) and the mean value of the output signal is consistently larger (or smaller) than the standard input, the difference between these signals is a bias-type error. The mean value added to the bias gives the standard input. Bias errors may be caused from temperature, off-ground voltage, reduced power supply, and so on. The source of the bias error should be known or the bias should be periodically checked to make sure that it does not change.

A second type of error that should be checked by calibration is the hysteresis error. This error occurs in some instruments when they indicate a value that is too low (or too high) when the inputs are increasing and too high (or too low) when the inputs are decreasing. A typical curve is shown in Figure AIII.1. If static calibration is first attempted with increasing input

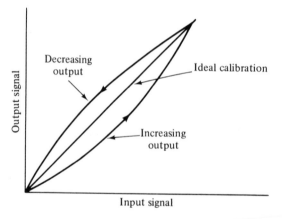

Figure AIII-1 Hysteresis Error

signals and then checked with decreasing input signals, a hysteresis error is observed when the output for one case is consistently higher than the output for the other case. For the measurement calibration to be used effectively, it is then also necessary to know whether the input signal is increasing or decreasing. Sources of this error are overdamped systems, thermal effects, loose motion, and wear.

For calibration to be most effective, all extraneous inputs should be known. Good electrical grounds, constant temperature and humidity conditions, and controlled power supplies are essential in maintaining the proper environment. However, when the measurement is moved to a different environment after calibration, the extraneous inputs should again be checked before the system is ready to measure an unknown signal. The interface between the sensor and the calibration system must also be checked in the measurement environment. For example, a thermocouple placed in an isothermal calibration bath may respond differently when placed in a nonisothermal environment. This interfacial problem is difficult to evaluate, and the resultant accuracy may be impossible to establish. However, if the problem is recognized, an approximate analysis of the measuring environment may be possible.

Static calibration is fundamental to all calibration processes. It establishes the static gain of a system to determine the constant, A_0, in the measurement system. To determine the other constants in Equation AIII.1, some form of dynamic calibration is necessary.

AIII.2
Dynamic Calibration

Dynamic calibration also requires a knowledge of the input signal. Here the input signal is to be a continuous function of time, and the source of basic dynamic standards is somewhat limited. The techniques to be used with a periodic input signal are covered in Chapter 5. In that chapter the periodic signal was a sine wave, but the techniques would be similar for more complex wave forms.

Four dynamic input signals that can be generated from standard static inputs are the step input, the ramp input, the terminated ramp input, and the impulse function. We have discussed the sine function in Chapter 5. These four types of input signals will be shown and discussed in the following pages. A mathematical treatment offers a better method of describing their uses than does a discussion of the physical principles.

The types of measurement systems in common use are either zeroth order $(A_1, A_2, \ldots, A_r = 0)$, first order $(A_2, A_3, \ldots, A_r = 0)$, or second order $(A_3, A_4, \ldots, A_r = 0)$. This is fortunate from the measurement point of view, since the solution of Equation AIII.1 becomes very involved with third-and higher-order systems. However, when the resonant frequencies of higher-order systems are widely separated, the techniques of Chapter 5 can be used to determine the coefficients in the governing equation. This will not be attempted here. Zeroth-order systems are calibrated by static calibrations. Third- and higher-order systems will not be discussed. Therefore, the response of first- and second-order systems to each of the four generated standard inputs will be considered here. Reference 2 should also be consulted for additional information. This is an excellent text on dynamic signals. Most of the material presented here is based on this reference.

FIRST-ORDER SYSTEMS

The basic differential equation to solve for a first-order system is

$$\left(\frac{A_1}{A_0}D + 1\right)I_o = \frac{1}{A_0}(B_m D^m + B_{m-1}D^{m-1} + \cdots + B_1 D + B_0)I_i \quad \text{(AIII.3)}$$

The constant A_1/A_0 is usually called the time constant, τ. A_0 will be assumed to be found from static calibration. The right-hand side of the equation will depend on which of the four input signals is selected. In all cases it must be a known value (preferably a signal fabricated from a known static standard).

For all cases of input signal the homogeneous solution of Equation AIII.3 is the same. It is found by solving

$$(\tau D + 1)I_o = 0$$
$$I_o = C_1 e^{-(t/\tau)} \quad \text{(AIII.4)}$$

The constant C_1 must be found from boundary conditions on the measurement system. A thermocouple, a thermistor, many other heat-transfer sensors, a charging capicator, the velocity of a free-falling mass and many other sensor systems may be approximated by first-order systems. The calibration method for determining τ could be relatively simple. (A thermocouple in air is suddenly immersed in ice water to obtain a dynamic step input signal.) The particular goal of this calibration is to obtain the value of τ. The mathematical techniques employed will be used to facilitate this goal.

The step input can be represented mathematically by the relationship

$$I_i = \begin{cases} 0 & \text{for } t < 0 \\ A & \text{for } t \geq 0 \end{cases}$$

This is shown in Figure AIII.2 along with the dymamic response of the first-order system. Using the above step input, Equation AIII.3 now becomes

$$(\tau D + 1)I_o = KA \qquad t \geq 0$$

where K is the static gain constant. The total solution is the sum of the homogenous solution and the particular solution for the step input signal,

$$I_o = KA(1 - e^{-t/\tau}) \quad \text{(AIII.5)}$$

It is possible to determine τ experimentally from the results of Figure AIII.2.

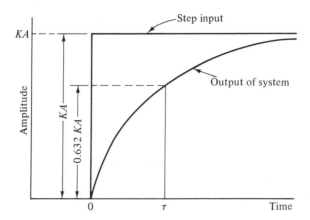

Figure AIII-2 First Order Response to a Step Input

When $t = \tau$, the term $1 - e^{-1} = 0.632$. This distance is measured off on the amplitude; the corresponding time is the desired time constant, τ. If the experiment is performed 20 times, a statistical analysis of τ and the corresponding standard deviation can be determined by the methods of Chapter 5. This type of statistical analysis is to be applied to all evaluation of constants for first- and second-order systems. Therefore, this will not be mentioned in the following discussion.

The step input provides one method of calibrating a first-order system. Voltages that vary linearly with time are also sources, for dynamic calibration. Mass flow, linear temperature gradients, and magnetic force fields can also exhibit linear variation with time. Therefore, a second signal source is a ramp input,

$$I_i = At \qquad t \geq 0$$

Equation AIII.3 for the ramp input is

$$(\tau D + 1)I_o = KAt$$

The particular solution is assumed in the form

$$I_{op} = C_2 t = C_3$$

Using this with the differential equation gives

$$\tau C_2 + C_2 t + C_3 = KAt$$
$$C_2 = KA \quad \text{and} \quad C_3 = -\tau AK$$

The general solution is the sum of this particular solution and the homogeneous solution,

$$I_o = KA(t - \tau) + C_1 e^{-t/\tau}$$

with the boundary condition when $t = 0$, $t_0 = 0$ this becomes

$$I_o = KA(t - \tau) + KA\tau e^{-t\tau}$$
$$KAt + KA\tau(e^{-t/\tau} - 1) \qquad \text{(AIII.6)}$$

The ramp input and the transient response of the first-order system to it are shown in Figure AIII.3. At steady-state conditions, when t is very large, $e^{-t/\tau}$ goes to zero, and

$$I_o = KA(t - \tau) = KI_i - KA\tau$$

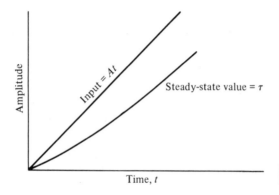

Figure AIII.3 First Order Response to a Ramp Input

The output signal at time equal to $t + \tau$ will have the same magnitude that the input had when the time was equal to t. This time constant can also be determined from the experimental data.

If the ramp input reaches a certain value, it may remain fixed at that value for certain input signals. This is called a terminated ramp input and is expressed mathematically in the form

$$I_i = At \quad \text{when} \quad 0 \le t \le T$$
$$= AT \quad \text{when} \quad t \ge T$$

Here the particular solution is considered in two parts, and the solutions are matched at time T. The first time interval has the same solution as that for the ramp input,

$$I_o = KAt + KA\tau(e^{-t/\tau} - 1) \qquad 0 \le t \le T \qquad \text{(AIII.7)}$$

The output for the terminated part has the same solution as that of the step input with a different boundary condition.

$$I_o = KAT + C_1 e^{-t/\tau}$$

with

$$I_o = KAT + KA\tau(e^{-t/\tau} - 1) \qquad t = T$$
$$C_1 = KA\tau(1 - e^{T/\tau})$$

or

$$I_o = KAT + KA\tau(e^{-t/\tau} - e^{T-t/\tau}) \qquad t \geq T \qquad \text{(AIII.8)}$$

This method is not normally used in calibration and is not shown. However, it may be necessary to use it in some particular case.

The response of piezoelectric crystals and strain gages is usually so high that a different form of input signal is required. One such input signal is the impulse-type signal. Here the input signal is ideally a sharp peak which occurs in zero time. The various forms of the impulse input are shown in Figure AIII.4. The real impulse occurs over a finite time and is assumed to be a constant for this time, T. Then,

$$I_i = \frac{A}{T} \quad \text{when} \quad 0 \leq t \leq T$$
$$= 0 \quad \text{when} \quad t > T$$

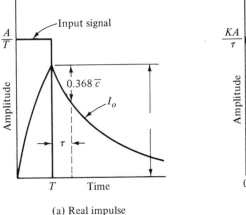

(a) Real impulse

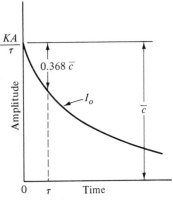

(b) Ideal impulse

Figure AIII.4 First Order Response to an Impulse

The solution to the first part is that of a step input,

$$I_o = \frac{KA}{T}(1 - e^{-t/\tau}) \qquad 0 \le t \le T \qquad \text{(AIII.9)}$$

And to the second part,

$$I_o = Ce^{-t/\tau} \qquad t > T$$

with a boundary match at $t = T$,

$$I_o = \frac{KA}{T}(e^{T/\tau} - 1)e^{-t/\tau} \qquad t > T \qquad \text{(AIII.10)}$$

In the limit as T goes to zero with the product $(A/T)(T)$ equal A, l'Hospital's rule gives

$$I_o = \frac{KA}{\tau} e^{-t/\tau}$$

Several methods are available for determining the time constant for this case. One such method is shown in Figure AIII.4. Although other methods may be used for dynamic calibration, we will limit our discussion to these. They will be adequate in most cases.

SECOND-ORDER SYSTEMS

The same input signals discussed above may also be used for second order systems. A system may have first-order response to some input frequency range and second-order response to other frequencies. The sine-wave input should also be checked in dynamic calibration. The signals discussed above can be used to classify a measurement system.

For a second-order system Equation AIII.1 becomes

$$\left(\frac{A_2}{A_0}D^2 + D + 1\right)I_o = \frac{1}{A_0}(B_m D^m + B_{m-1}D^{m-1} + \cdots + B_1 D + D_0)I_i$$

$$\text{(AIII.11)}$$

The homogeneous solution of this equation is expressed using the more common notation of classical vibration theory. For this notation the natural frequency of the system is

$$\omega_n = \frac{\sqrt{A_0}}{A_2} \quad \text{and} \quad \zeta = \frac{A_1}{2\sqrt{A_0 A_2}}$$

The homogeneous equation becomes

$$\left(\frac{D^2}{\omega_n^2} + \frac{2\zeta D}{\omega_n} + 1\right) I_o = 0$$

The complimentary solution depends on the sign of the roots of the associated algebraic equation

$$r = -\zeta\omega_n \pm \omega_n\sqrt{\zeta^2 - 1}$$

If $\zeta > 1$, the solution is called overdamped, and

$$I_o = C_1 e^{(-\zeta + \sqrt{\zeta^2 - 1})\omega_n t} + C_2 e^{(-\zeta - \sqrt{\zeta^2 - 1})\omega_n t} \qquad \text{(AIII.12)}$$

If $\zeta = 1$, the solution is called critically damped, and

$$I_o = C_1 e^{-\omega_n t} + C_2 t e^{-\omega_n t} \qquad \text{(AIII.13)}$$

If $\zeta = 1$, the solution is called underdamped, and

$$I_o = e^{-\zeta\omega_n t}(C_1 \cos \sqrt{1 - \zeta^2}\, \omega_n t + C_2 \sin \sqrt{1 - \zeta^2}\, \omega_n t) \qquad \text{(AIII.14)}$$

The space required to consider three cases for each input signal is too large. Therefore, only the step input and the impulse input will be presented here. Some comments will be made concerning the sine-wave input for the second-order system will also be made.

For the step input Equation AIII.11 becomes

$$\left(\frac{D^2}{\omega_n^2} + \frac{2\zeta D}{\omega_n} + 1\right) I_o = KA \qquad t \geq 0$$

The general solution for the step input with an overdamped system is

$$I_o = KA + C_1 \exp\left[(-\zeta + \sqrt{\zeta^2 - 1})(\omega_n t)\right] + C_2 \exp\left[(-\zeta - \sqrt{\zeta^2 - 1})(\omega_n t)\right]$$

If the output and its derivative is zero at $t = 0$, the constants C_1 and C_2 can be evaluated by solving two equations in two unknowns:

$$I_o = KA \left\{ 1 - \frac{-\zeta + \sqrt{\zeta^2 - 1}}{2\sqrt{\zeta^2 - 1}} \exp\left[(-\zeta + \sqrt{\zeta^2 - 1})(\omega_n t)\right] \right.$$
$$\left. + \frac{-\zeta - \sqrt{\zeta^2 - 1}}{2\sqrt{\zeta^2 - 1}} \exp\left[(-\zeta - \sqrt{\zeta^2 - 1})(\omega_n t)\right] \right\} \qquad \text{(AIII.15)}$$

It is actually very difficult to use a step input signal to determine the measurement system constants when the system is either overdamped or critically damped. The mathematical results are presented here for completeness, but the sine wave is the best method for determining ω_n and ζ for these cases. The solution for the critically damped case is

$$I_o = KA[1 - (1 + \omega_n t)e^{-\omega_n t}] \tag{AIII.16}$$

When the system is underdamped, the output is

$$I_o = KA\left[1 - \frac{1}{\sqrt{1 - \zeta^2}} e^{-\zeta\omega_n t} \sin\left(\sqrt{1 - \zeta^2}\,\omega_n t + \phi\right)\right]$$

where

$$\phi = \sin^{-1}\sqrt{1 - \zeta^2}$$

All three cases are shown in Figure AIII.5. It is possible to obtain the constants ω_n and ζ in the following manner:

1. The measurement of time and peak positive amplitude for two peaks can be used to obtain the product $\zeta\omega_n$ by determining the slope of the exponential term.

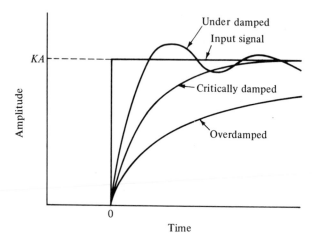

Figure AIII.5 Second Order Response to a Step Input

2. The time between peaks determines the period of oscillation and, therefore, the term $\sqrt{1 - \zeta^2}\,\omega_n$. The two equations can be solved simultaneously to obtain ω_n and ζ. The constants for the system can then be determined. (Remember that A_0 was found by static calibration.)

The dynamic response of a second-order system to an ideal impulse input can also be used to obtain information for the underdamped case. Again the information is difficult to obtain for the overdamped and the critically damped cases. If the boundary conditions for an ideal impulse signal (time interval of the impulse is zero) are $I_o = 0$ and $dI_o/dt = KA\omega_n^2$ at $t = 0$, reference 2 gives the results of the output signal for the overdamped case,

$$I_o = \frac{KA\omega_n}{2\sqrt{\zeta^2 - 1}} \{\exp[(-\zeta + \sqrt{\zeta^2 - 1})(\omega_n t)] - \exp[(-\zeta - \sqrt{\zeta^2 - 1})(\omega_n t)]\}$$

$$\text{(AIII.17)}$$

for the critically damped case,

$$I_o = KA\omega_n^2 t e^{-\omega_n t} \qquad \text{(AIII.18)}$$

and for the underdamped case,

$$I_o = \frac{KA\omega_n}{\sqrt{1 - \zeta^2}} e^{-\zeta\omega_n t} \sin(\sqrt{1 - \zeta^2}\,\omega_n t) \qquad \text{(AIII.19)}$$

Figure AIII.6 was taken from this source to show the dynamic characteristics of a second-order system to an impulse input.

One method of finding the constants of a second-order system that is critically damped or overdamped is to use the response curves of Figure

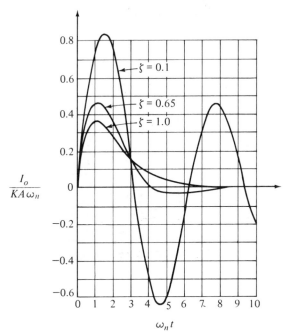

$\dfrac{I_o}{KA\omega_n}$

$\omega_n t$

Figure AIII.6 Second Order Response to an Impulse Input

AIII.7. These were obtained in reference 2 for a sine-wave input signal. The natural frequency of the system can be obtained when the phase lag is 90°. The damping can be approximated from the dimensionless amplitude ratio at the natural frequency. This method is still not as accurate as we desire, but it can approximate and bound the constants. Since most measurement systems have damping ratios of 0.7 (actual damping divided by

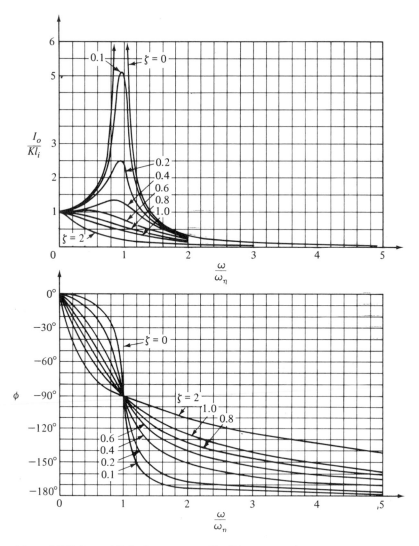

Figure AIII.7 Second Order Response to a Sine Wave

critical damping), these curves could be used in many cases. These methods can be very valuable because the available choices are limited.

Finally, it should be mentioned that all dynamic calibration involves one form of error signal. The time required for an input signal to travel from the source to the sensor could be important in many situations. This time is an unavoidable delay in transmission. The time required for a signal to go from the source to sensor is called dead time. It can be evaluated when this value is necessary. In many cases a pure dead-time element is not a major problem. Then the dead time need not be determined.

Dynamic calibration for higher-order systems becomes even more complex than the methods discussed here. In most sensors the higher-order derivatives are small enough to be neglected. If this is not possible, some combination of the techniques presented in this appendix might be used. Fortunately, almost all measurement systems are of order 2 or less, and the methods presented here can give some measure of the calibration.

One powerful tool used in analyzing the output of a measurement system for sine wave inputs is the Fourier integral method for spectral analysis. Details of this technique are given in references 3 and 4. This method will not be discussed here, but it is important that second-order systems have available the best techniques for calibration. The Fourier analysis can be used in situations where other simpler methods fail.

References

1. DANEMAN, H. L., *Some Suggestions Toward Equipping a Standard Laboratory*, Leeds and Northrup Co. Technical Publication A 0.0111 2nd ed., 1969.

2. DOEBELIN, E. O., *Measurement Systems: Application and Design*, McGraw-Hill Book Co., Inc., New York, 1966.

3. SCHWEPPE, J. L., EICHBERGER, L. C., MUSTER, D. F., MICHAELS, E. L., and PASKUSZ, G. F., *Methods for the Dynamic Calibration of Pressure Transducers*, NBS Monograph 67, U. S. Department of Commerce, Washington, D.C., 1963.

4. DANIEL, R. G., "Experimental Frequency Response Determination," *Instruments and Control Systems*, June 1969, p. 113.

INDEX

A

Absolute error limts, 105-108, 135
Acceleration:
  angular, 274-75
  linear, 270-74
Accuracy:
  angles, 222
  of measurement, 73-75, 105, 221
  percent, 75
  of sensors, 32-34
Accuracy and range, 28
Alphatron, 306-307
Ammeter, 45
Amplitude, 110
Analog model, 11, 16, 59
Analysis of program, 2-3, 105, 135
Angular measurement, 221-23
Angular velocity, 268-70
Area, 223-26
Atomic clock, 167
Average measured value, 74, 83-86

B

Ballast circuit, 156-57
Bar graph, 78, 89
Bathroom scales, 235
Beam balance, 164-66
Bias error, 75-77, 80
Bimetal sensor, 190-91
Bourdon tube, 255-56
Buckingham Pi Theorem, 15, 332
Burrette, 230-31

C

Calibration:
  analysis, 102

Calibration: (*Contd.*)
  angles, 221
  dynamic, 52, 343-53
  fundamentals, 340-53
  length, 142-43
  mass, 164
  resistance, 212-14
  static, 52, 340-43
  temperature, 174-76, 188
  time, 167-71
  voltage, 203-04
    ac, 210
Caliper, 146-48
Calorimeter, 296-97
Capacitance bridges, 283-84
Cathode-ray tube, 204-205
Cesium *133*, 167
Chi-squared test, 88-93
Colorimeter, 319-21
Control parameters, 12-13, 63
Curve fitting, 103-104

D

D'Arsonval meter, 44, 197-99
Data analysis, 23, 71-135
Dead-weight tester, 254-55
Decibel, 112
Deflection balance, 233
Density, 229-32
Design of experiment, 2-3, 9-12, 63-66, 135
Design of measurement system, 2-3, 23-24, 66, 135
Deviation, 86
Dial indicator, 146-48
Diaphragm gage, 256-57
Differential transformer, 162-63
Digital computer modeling, 11, 17